Common Core, Inc. (commoncore.org) is a non-profit organization formed in 2007 to advocate for a content-rich liberal arts education in America's K-12 schools. To improve education in America, we create curriculum materials, conduct professional development, and also promote programs, policies, and initiatives at the local, state, and federal levels that provide students with challenging, rigorous instruction in the full range of liberal arts and sciences.

Common Core, Inc. is not affiliated with the Common Core State Standards Initiative.

Special thanks go to the Gordan A. Cain Center and to the Department of Mathematics at Louisiana State University for their support in the development of *Eureka Math*.

Published by Common Core

Common Core
1016 16th Street NW, 7th Floor
Washington, DC 20036
Phone 202.223.1854
Web commoncore.org
Email info@commoncore.org

Printed in the U.S.A.
This book may be purchased from the publisher at commoncore.org
10 9 8 7 6 5 4 3 2 1

ISBN 978-1-63255-028-6

Name ________________________________ Date ________________

1. Determine the perimeter and area of rectangles A and B.

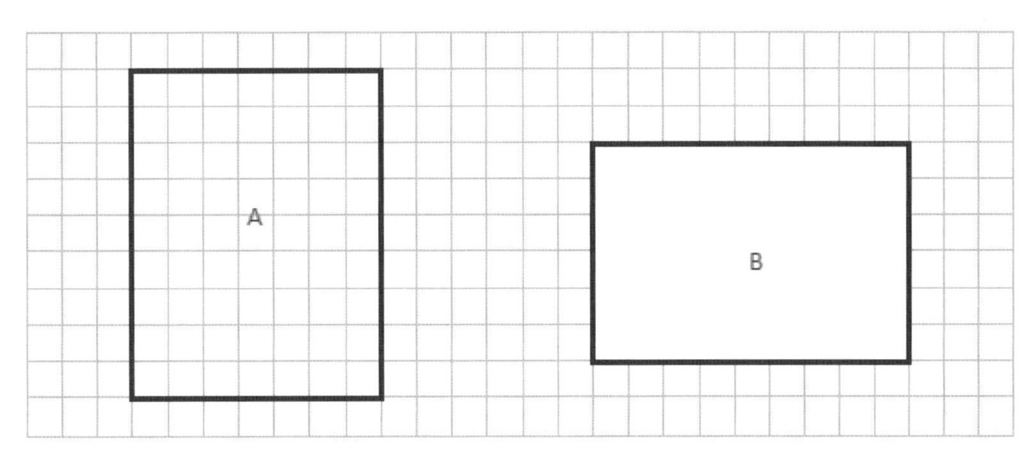

A = ____________ A = ____________

P = ____________ P = ____________

2. Determine the perimeter and area of each rectangle.

a.

6 cm

5 cm

P = ____________

A = ____________

b.

3 cm

8 cm

P = ____________

A = ____________

3. Determine the perimeter of each rectangle.

a.

P = ____________

b.

P = ____________

4. Given the rectangle's area, find the unknown side length.

a.

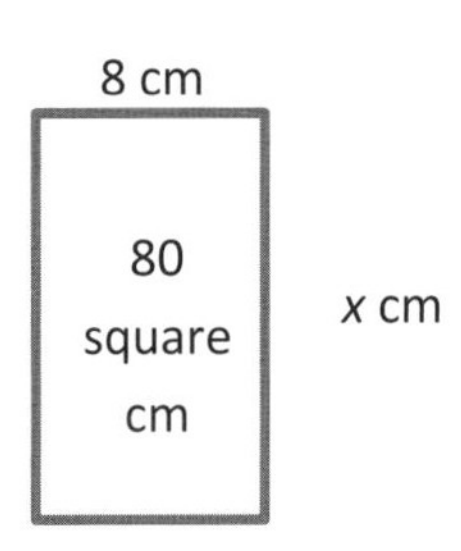

x = ____________

b.

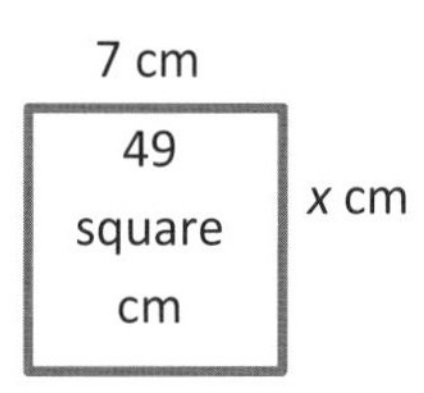

x = ____________

5. Given the rectangle's perimeter, find the unknown side length.

a. P = 120 cm

20 cm

x cm

x = _______________

b. P = 1,000 m

x m

250 m

x = _______________

6. Each of the following rectangles has whole number side lengths. Given the area and perimeter, find the length and width.

a. P = 20 cm

l = _________

24 square cm

w = _______

b. P = 28 m

w = _______

24 square m

l = _________

Name ______________________________ Date ______________

1. Determine the perimeter and area of rectangles A and B.

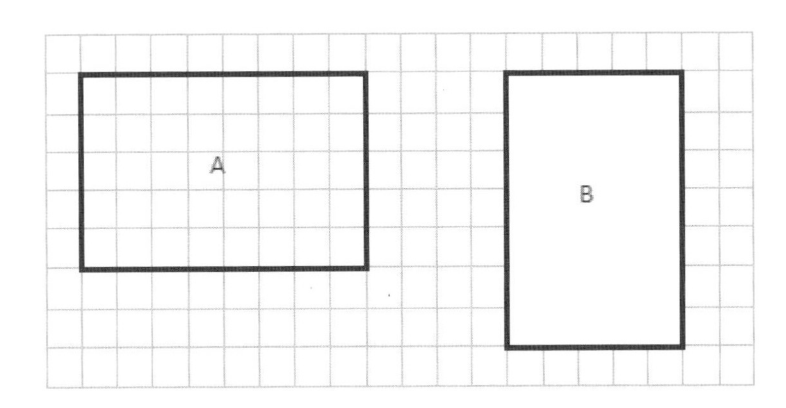

A = ______________ A = ______________

P = ______________ P = ______________

2. Determine the perimeter and area of each rectangle.

a.

7 cm

3 cm

P = ______________

A = ______________

b.

84 cm

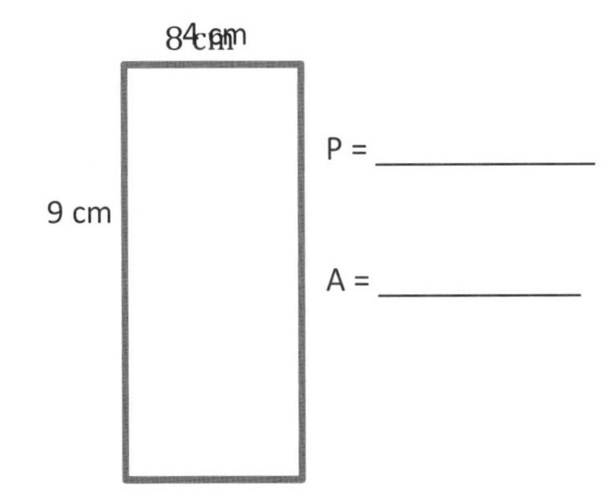

9 cm

P = ______________

A = ______________

3. Determine the perimeter of each rectangle.

a.

149 m

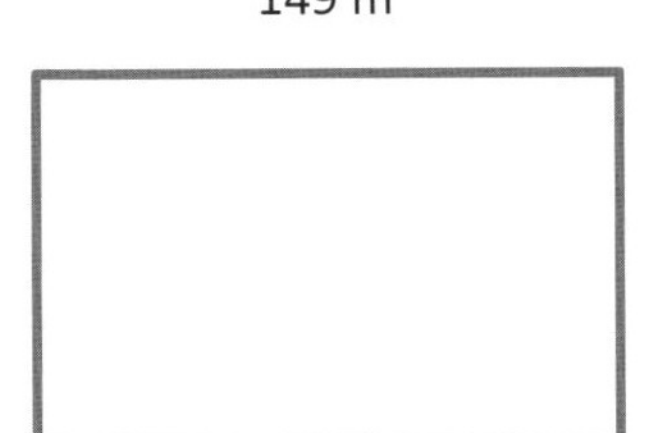

76 m

P = ______________

b.

2 m 10 cm

45 cm

P = ______________

4. Given the rectangle's area, find the unknown side length.

a.

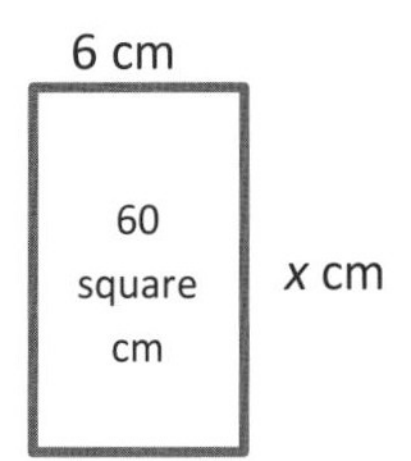

x = ____________

b.

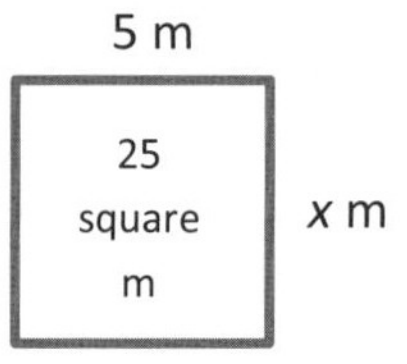

x = ____________

5. Given the rectangle's perimeter, find the unknown side length.

a. P = 180 cm 40 cm

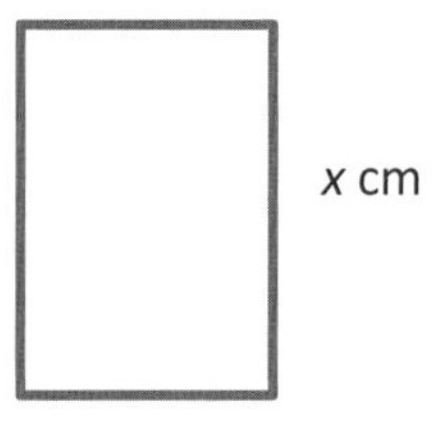

x = _______________

b. P = 1,000 m

x m

150 m

x = _______________

6. Each of the following rectangles has whole number side lengths. Given the area and perimeter, find the length and width.

a. A = 32 square cm
P = 24 cm

l = __________

32 square cm

w = _______

b. A = 36 square m
P = 30 m

w = _______

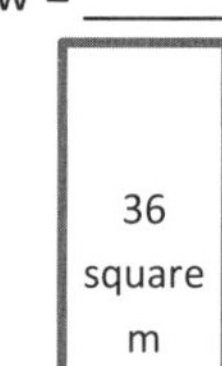

l = _________

Name ______________________________ Date ______________

1. A rectangular porch is 4 feet wide. It is 3 times as long as it is wide.

   a. Label the diagram with the dimensions of the porch.

   b. Find the perimeter of the porch.

2. A narrow rectangular banner is 5 inches wide. It is 6 times as long as it is wide.

   a. Draw a diagram of the banner and label its dimensions.

   b. Find the perimeter and area of the banner.

3. The area of a rectangle is 42 square centimeters. Its length is 7 centimeters.

   a. What is the width of the rectangle?

   b. Charlie wants to draw a second rectangle that is the same length but is 3 times as wide. Draw and label Charlie's second rectangle.

   c. What is the perimeter of Charlie's second rectangle?

4. The area of Betsy's rectangular sandbox is 20 square feet. The longer side measures 5 feet. The sandbox at the park is twice as long and twice as wide as Betsy's.

   a. Draw and label a diagram of Betsy's sandbox. What is its perimeter?

   b. Draw and label a diagram of the sandbox at the park. What is its perimeter?

c. What is the relationship between the two perimeters?

d. Find the area of the park's sandbox using the formula $A = l \times w$.

e. The sandbox at the park has an area that is how many times that of Betsy's sandbox?

f. Compare how the perimeter changed with how the area changed between the two sandboxes. Explain what you notice using words, pictures, or numbers.

Name ______________________________ Date ________________

1. A rectangular pool is 7 feet wide. It is 3 times as long as it is wide.

   a. Label the diagram with the dimensions of the pool.

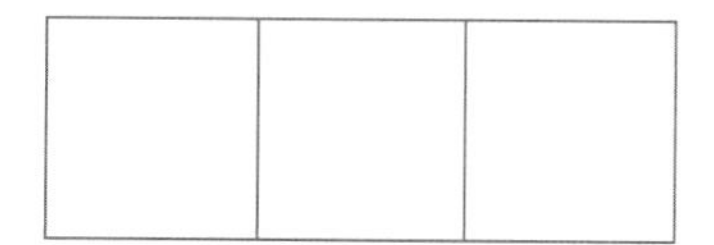

   b. Find the perimeter of the pool.

2. A poster is 3 inches long. It is 4 times as wide as it is long.

   a. Draw a diagram of the poster and label its dimensions.

   b. Find the perimeter and area of the poster.

3. The area of a rectangle is 36 square centimeters and its length is 9 centimeters.

   a. What is the width of the rectangle?

   b. Elsa wants to draw a second rectangle that is the same length but is 3 times as wide. Draw and label Elsa's second rectangle.

   c. What is the perimeter of Elsa's second rectangle?

4. The area of Nathan's bedroom rug is 15 square feet. The longer side measures 5 feet. His living room rug is twice as long and twice as wide as the bedroom rug.

   a. Draw and label a diagram of Nathan's bedroom rug. What is its perimeter?

   b. Draw and label a diagram of Nathan's living room rug. What is its perimeter?

c. What is the relationship between the two perimeters?

d. Find the area of the living room rug using the formula $A = l \times w$.

e. The living room rug has an area that is how many times that of the bedroom rug?

f. Compare how the perimeter changed with how the area changed between the two rugs. Explain what you notice using words, pictures, or numbers.

Name ______________________________ Date ________________

Solve the following problems. Use pictures, numbers, or words to show your work.

1. The rectangular projection screen in the school auditorium is 5 times as long and 5 times as wide as the rectangular screen in the library. The screen in the library is 4 feet long with a perimeter of 14 feet. What is the perimeter of the screen in the auditorium?

2. The width of David's rectangular tent is 5 feet. The length is twice the width. David's rectangular air mattress measures 3 feet by 6 feet. If David puts the air mattress in the tent, how many square feet of floor space will be available for the rest of his things?

3. Jackson's rectangular bedroom has an area of 90 square feet. The area of his bedroom is 9 times that of his rectangular closet. If the closet is 2 feet wide, what is its length?

4. The length of a rectangular deck is 4 times its width. If the deck's perimeter is 30 feet, what is the deck's area?

Name ______________________________ Date ________________

Solve the following problems. Use pictures, numbers, or words to show your work.

1. Katie cut out a rectangular piece of wrapping paper that was 2 times as long and 3 times as wide as the box that she was wrapping. The box was 5 inches long and 4 inches wide. What is the perimeter of the wrapping paper that Katie cut?

2. Alexis has a rectangular piece of red paper that is 4 centimeters wide. Its length is twice its width. She glues a rectangular piece of blue paper on top of the red piece measuring 3 centimeters by 7 centimeters. How many square centimeters of red paper will be visible on top?

3. Brinn's rectangular kitchen has an area of 81 square feet. The kitchen is 9 times as many square feet as Brinn's pantry. If the rectangular pantry is 3 feet wide, what is the length of the pantry?

4. The length of Marshall's rectangular poster is 2 times its width. If the perimeter is 24 inches, what is the area of the poster?

Name ____________________ Date ____________

Example:

5 × 10 = 50

5 ones × 10 = 5 tens

thousands	hundreds	tens	ones
		(o o o o o) ← ×10	(● ● ● ● ●)

Draw place value disks and arrows as shown to represent each product.

1. 5 × 100 = ________

   5 × 10 × 10 = ________

   5 ones × 100 = ____ ________

thousands	hundreds	tens	ones

2. 5 × 1,000 = ________

   5 × 10 × 10 × 10 = ________

   5 ones × 1,000 = ____ ________

thousands	hundreds	tens	ones

3. Fill in the blanks in the following equations.

   a. 6 × 10 = ______ b. ______ × 6 = 600 c. 6,000 = ______ × 1,000

   d. 10 × 4 = ______ e. 4 × ______ = 400 f. ______ × 4 = 4,000

   g. 1,000 × 9 = ______ h. ______ = 10 × 9 i. 900 = ______ × 100

Draw place value disks and arrows to represent each product.

4. 12 × 10 = __________

(1 ten 2 ones) × 10 = _______________

thousands	hundreds	tens	ones

5. 18 × 100 = __________

18 × 10 × 10 = __________

(1 ten 8 ones) × 100 = ________________

thousands	hundreds	tens	ones

6. 25 × 1,000 = __________

25 × 10 × 10 × 10 = __________

(2 tens 5 ones) × 1,000 = ________________

ten thousands	thousands	hundreds	tens	ones

Decompose each multiple of 10, 100, or 1,000 before multiplying.

7. 3 × 40 = 3 × 4 × _____

= 12 × ______

= __________

8. 3 × 200 = 3 × _____ × ______

= ______ × ______

= ________

9. 4 × 4,000 = _____ × _____ × _________

= ______ × _________

= _________

10. 5 × 4,000 = _____ × _____ × _________

= ______ × ________

= ______

Name ______________________________ Date ______________

Example:

5 × 10 = 50

5 ones × 10 = 5 tens

thousands	hundreds	tens	ones
		○○○○○ (×10)	●●●●●

Draw place value disks and arrows as shown to represent each product.

1. 7 × 100 = ________

   7 × 10 × 10 = ________

   7 ones × 100 = ____ __________

thousands	hundreds	tens	ones

2. 7 × 1,000 = ________

   7 × 10 × 10 × 10 = ________

   7 ones × 1,000 = ____

   ____

thousands	hundreds	tens	ones

3. Fill in the blanks in the following equations.

   a. 8 × 10 = ________

   b. ______ × 8 = 800

   c. 8,000 = ______ × 1,000

   d. 10 × 3 = ______

   e. 3 × ______ = 3,000

   f. ______ × 3 = 300

   g. 1,000 × 4 = ______

   h. ______ = 10 × 4

   i. 400 = ______ × 100

Draw place value disks and arrows to represent each product.

4. 15 × 10 = ________

(1 ten 5 ones) × 10 = ____ ________

thousands	hundreds	tens	ones

5. 17 × 100 = ________

17 × 10 × 10 = ________

(1 ten 7 ones) × 100 = ____ ________

thousands	hundreds	tens	ones

6. 36 × 1,000 = ________

36 × 10 × 10 × 10 = ________

(3 tens 6 ones) × 1,000 = ____ ________

ten thousands	thousands	hundreds	tens	ones

Decompose each multiple of 10, 100, or 1000 before multiplying.

7. 2 × 80 = 2 × 8 × _____

= 16 × _____

= ________

8. 2 × 400 = 2 × _____ × _____

= _____ × _____

= ________

9. 5 × 5,000 = _____ × _____ × ________

= _____ × ________

= ________

10. 7 × 6,000 = _____ × _____ × ________

= _____ × ________

= _____

thousands	hundreds	tens	ones

thousands place value chart

Name ________________________________ Date ________________

Draw place value disks to represent the value of the following expressions.

1. $2 \times 3 =$ ______

   2 times _____ ones is _____ ones.

thousands	hundreds	tens	ones

$$\begin{array}{r} 3 \\ \times\ 2 \\ \hline \end{array}$$

2. $2 \times 30 =$ ______

   2 times _____ tens is _____ __________.

thousands	hundreds	tens	ones

$$\begin{array}{r} 30 \\ \times\ \ 2 \\ \hline \end{array}$$

3. $2 \times 300 =$ ______

   2 times ______ __________ is ______ ____________ .

thousands	hundreds	tens	ones

$$\begin{array}{r} 300 \\ \times\ \ \ 2 \\ \hline \end{array}$$

4. $2 \times 3{,}000 =$ ______

   ____ times ________________ is ________________ .

thousands	hundreds	tens	ones

$$\begin{array}{r} 3{,}000 \\ \times\ \ \ \ 2 \\ \hline \end{array}$$

5. Find the product.

a. 20×7	b. 3×60	c. 3×400	d. 2×800
e. 7×30	f. 60×6	g. 400×4	h. $4 \times 8{,}000$
i. 5×30	j. 5×60	k. 5×400	l. $8{,}000 \times 5$

6. Brianna buys 3 packs of balloons for a party. Each pack has 60 balloons. How many balloons does Brianna have?

7. Jordan has twenty times as many baseball cards as his brother. His brother has 9 cards. How many cards does Jordan have?

8. The aquarium has 30 times as many fish in one tank as Jacob has. The aquarium has 90 fish. How many fish does Jacob have?

Name ______________________________ Date ______________

Draw place value disks to represent the value of the following expressions.

1. 5 × 2 = ______

   5 times _____ ones is _____ ones.

thousands	hundreds	tens	ones

$$\begin{array}{r} 2 \\ \times\ 5 \\ \hline \end{array}$$

2. 5 × 20 = ______

   5 times _____ tens is _______________.

thousands	hundreds	tens	ones

$$\begin{array}{r} 20 \\ \times\ \ 5 \\ \hline \end{array}$$

3. 5 × 200 = ______

   5 times _______ ____________ is ______ ______________ .

thousands	hundreds	tens	ones

$$\begin{array}{r} 200 \\ \times\ \ \ 5 \\ \hline \end{array}$$

4. 5 × 2,000 = ______

   ____ times _______ ______________ is _______ _______________ .

thousands	hundreds	tens	ones

$$\begin{array}{r} 2{,}000 \\ \times\ \ \ \ 5 \\ \hline \end{array}$$

5. Find the product.

a. 20 × 9	b. 6 × 70	c. 7 × 700	d. 3 × 900
e. 9 × 90	f. 40 × 7	g. 600 × 6	h. 8 × 6,000
i. 5 × 70	j. 5 × 80	k. 5 × 200	l. 6,000 × 5

6. At the school cafeteria, each student who ordered lunch gets 6 chicken nuggets. The cafeteria staff prepares enough for 300 kids. How many chicken nuggets does the cafeteria staff prepare altogether?

7. Jaelynn has 30 times as many stickers as her brother. Her brother has 8 stickers. How many stickers does Jaelynn have?

8. The flower shop has 40 times as many flowers in one cooler as Julia has in her bouquet. The cooler has 120 flowers. How many flowers are in Julia's bouquet?

Name ______________________________ Date ________________

Represent the following problem by drawing disks in the place value chart.

1. To solve 20×40, think:

   $(2 \text{ tens} \times 4) \times 10 =$ ________

   $20 \times (4 \times 10) =$ ________

   $20 \times 40 =$ _______

hundreds	tens	ones

2. Draw an area model to represent 20×40.

   $2 \text{ tens} \times 4 \text{ tens} =$ _____ __________

3. Draw an area model to represent 30×40.

   $3 \text{ tens} \times 4 \text{ tens} =$ _____ __________

   $30 \times 40 =$ ______

4. Draw an area model to represent 20 × 50.

2 tens × 5 tens = _____ ___________

20 × 50 = _______

Rewrite each equation in unit form and solve.

5. 20 × 20 = ________

2 tens × 2 tens = _____ hundreds

6. 60 × 20 = _______

6 tens × 2 ________ = ____ hundreds

7. 70 × 20 = _______

_____ tens × _____ tens = 14 _________

8. 70 × 30 = _______

____ _______ × ____ _______ = _____ hundreds

9. If there are 40 seats per row, how many seats are in 90 rows?

10. One ticket to the symphony costs $50. How much money is collected if 80 tickets are sold?

Name ______________________________ Date ______________

Represent the following problem by drawing disks in the place value chart.

1. To solve 30 × 60, think:

   (3 tens × 6) × 10 = ________

   30 × (6 × 10) = ________

   30 × 60 = _______

hundreds	tens	ones

2. Draw an area model to represent 30 × 60.

   3 tens × 6 tens = _____ ____________

3. Draw an area model to represent 20 × 20.

   2 tens × 2 tens = _____ ____________

   20 × 20 = ______

4. Draw an area model to represent 40×60.

4 tens × 6 tens = _____ ____________

40×60 = _______

Rewrite each equation in unit form and solve.

5. 50×20 = ________

5 tens × 2 tens = _____ hundreds

6. 30×50 =

3 tens × 5 ________ = ____ hundreds

7. 60×20 =

_____ tens × _____ tens = 12 _________

8. 40×70 =

____ _______ × ____ _______ = _____ hundreds

9. There are 60 seconds in a minute and 60 minutes in an hour. How many seconds are in one hour?

10. To print a comic book, 50 pieces of paper are needed. How many pieces of paper are needed to print 40 comic books?

Name ______________________________ Date ______________

1. Represent the following expressions with disks, regrouping as necessary, writing a matching expression, and recording the partial products vertically as shown below.

   a. 1×43

tens	ones
● ● ● ●	● ● ●

1×4 tens + 1×3 ones

$$\begin{array}{rr} & 4\ 3 \\ \times & 1 \\ \hline & 3 \\ + & 4\ 0 \\ \hline & 4\ 3 \end{array}$$

3 → 1×3 ones

40 → 1×4 tens

   b. 2×43

tens	ones

   c. 3×43

hundreds	tens	ones

d. 4×43

hundreds	tens	ones

2. Represent the following expressions with disks, regrouping as necessary. To the right, record the partial products vertically.

a. 2×36

hundreds	tens	ones

b. 3×61

hundreds	tens	ones

c. 4×84

hundreds	tens	ones

Name ______________________________ Date ______________

1. Represent the following expressions with disks, regrouping as necessary, writing a matching expression, and recording the partial products vertically.

a. 3 × 24

tens	ones

b. 3 × 42

hundreds	tens	ones

c. 4 × 34

hundreds	tens	ones

2. Represent the following expressions with disks, regrouping as necessary. To the right, record the partial products vertically.

   a. 4 × 27

hundreds	tens	ones

   b. 5 × 42

hundreds	tens	ones

3. Cindy says she found a shortcut for doing multiplication problems. When she multiplies 3 × 24, she says, "3 × 4 is 12 ones, or 1 ten and 2 ones. Then, there's just 2 tens left in 24, so add it up, and you get 3 tens and 2 ones." Do you think Cindy's shortcut works? Explain your thinking in words and justify your response using a model or partial products.

ten thousands	thousands	hundreds	tens	ones

ten thousands place value chart

Name ______________________________ Date ______________

1. Represent the following expressions with disks, regrouping as necessary, writing a matching expression, and recording the partial products vertically as shown below.

a. 1 × 213

hundreds	tens	ones

```
    2  1  3
×         1
-----------
             → 1 × 3 ones
             → 1 × 1 ten
+            → 1 × 2 hundreds
-----------
```

1 × ___ hundreds + 1 × ___ ten + 1 × ___ ones

b. 2 × 213

hundreds	tens	ones

c. 3 × 214

hundreds	tens	ones

d. $3 \times 1,254$

thousands	hundreds	tens	ones

2. Represent the following expressions with disks, using either method shown during class, regrouping as necessary. To the right, record the partial products vertically.

a. 3×212

b. $2 \times 4,036$

c. $3 \times 2,546$

d. $3 \times 1,407$

3. Every day at the bagel factory, Cyndi makes 5 different kinds of bagels. If she makes 144 of each kind, what is the total number of bagels that she makes?

Name ______________________________ Date ______________

1. Represent the following expressions with disks, regrouping as necessary, writing a matching expression, and recording the partial products vertically as shown below.

a. 2 × 424

hundreds	tens	ones
● ● ● ●	● ●	● ● ● ●

```
    4  2  4
×         2
-----------
              → 2 × ___ ones
              → 2 × ___ ______
+             → ___ × ___ ________
-----------
```

2 × ___ ________ + 2 × ___ ______ + 2 × ___ ones

b. 3 × 424

hundreds	tens	ones

c. 4 × 1,424

2. Represent the following expressions with disks, using either method shown in class, regrouping as necessary. To the right, record the partial products vertically.

   a. 2×617

   b. 5×642

   c. $3 \times 3{,}034$

3. Every day, Penelope jogs three laps around the playground to keep in shape. The playground is rectangular with a width of 163 m and a length of 320 m.

   a. Find the total amount of meters in one lap.

   b. Determine how many meters Penelope jogs in three laps.

Name ______________________________ Date ________________

1. Solve using each method.

Partial Products	Standard Algorithm
a. 3 4 × 4	3 4 × 4

Partial Products	Standard Algorithm
b. 2 2 4 × 3	2 2 4 × 3

2. Solve. Use the standard algorithm.

a. 2 5 1 × 3	b. 1 3 5 × 6	c. 3 0 4 × 9
d. 4 0 5 × 4	e. 3 1 6 × 5	f. 3 9 2 × 6

3. The product of 7 and 86 is ________.

4. 9 times as many as 457 is _________.

5. Jashawn wants to make 5 airplane propellers.
He needs 18 centimeters of wood for each propeller.
How many centimeters of wood will he use?

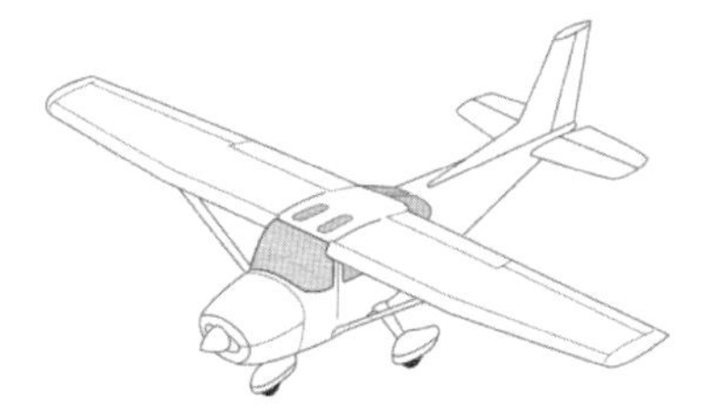

6. One game system costs $238. How much will 4 game systems cost?

7. A small bag of chips weighs 48 grams. A large bag of chips weighs three times as much as the small bag. How much will 7 large bags of chips weigh?

Name ______________________________　　Date ______________

1. Solve using each method.

Partial Products	Standard Algorithm
a. 46×2	46×2

Partial Products	Standard Algorithm
b. 315×4	315×4

2. Solve using the standard algorithm.

a. 232×4	b. 142×6	c. 314×7
d. 440×3	e. 507×8	f. 384×9

3. What is the product of 8 and 54?

4. Isabel earned 350 points while she was playing Blasting Robot. Isabel's mom earned 3 times as many points as Isabel. How many points did Isabel's mom earn?

5. To get enough money to go to on a field trip, every student in a club has to raise $53 by selling chocolate bars. There are 9 students in the club. How much money does the club need to raise to go on the field trip?

6. Mr. Meyers wants to order 4 tablets for his classroom. Each tablet costs $329. How much will all four tablets cost?

7. Amaya read 64 pages last week. Amaya's older brother, Rogelio, read twice as many pages in the same amount of time. Their big sister, Elianna, is in high school and read 4 times as many pages as Rogelio did. How many pages did Elianna read last week?

Name ______________________ Date ____________

1. Solve using the standard algorithm.

a. 3×42	b. 6×42
c. 6×431	d. 3×431
e. $3 \times 6{,}212$	f. $3 \times 3{,}106$
g. $4 \times 4{,}309$	h. $4 \times 8{,}618$

2. There are 365 days in a common year. How many days are in 3 common years?

3. The length of one side of a square city block is 462 meters. What is the perimeter of the block?

4. Jake ran 2 miles. Jesse ran 4 times as far. There are 5,280 feet in a mile. How many feet did Jesse run?

Name ______________________________ Date ______________

1. Solve using the standard algorithm.

a. 3×41	b. 9×41
c. 7×143	d. 7×286
e. $4 \times 2,048$	f. $4 \times 4,096$
g. $8 \times 4,096$	h. $4 \times 8,192$

2. Robert's family brings six gallons of water for the players on the football team. If one gallon of water contains 128 fluid ounces, how many fluid ounces are in six gallons?

3. It takes 687 Earth days for the planet Mars to revolve around the Sun once. How many Earth days does it take Mars to revolve around the Sun four times?

4. Tammy buys a 4-gigabyte memory card for her camera. Dijonea buys a memory card with twice as much storage as Tammy's. One gigabyte is 1,024 megabytes. How many megabytes of storage does Dijonea have on her memory card?

Name ______________________________ Date ______________

1. Solve the following expressions using the standard algorithm, the partial products method, and the area model.

a. 4 2 5 × 4

4 (400 + 20 + 5)

(4 × _____) + (4 × _____) + (4 × _____)

b. 5 3 4 × 7

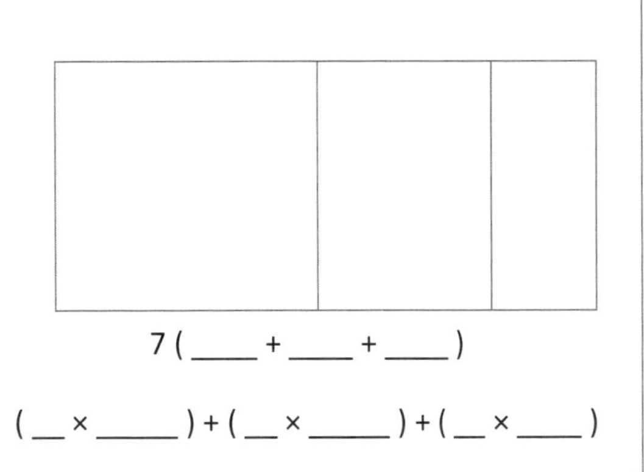

7 (____ + ____ + ____)

(__ × _____) + (__ × _____) + (__ × _____)

c. 2 0 9 × 8

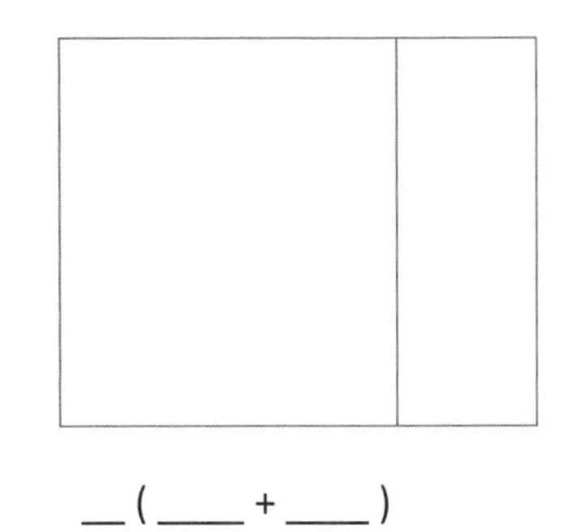

__ (____ + ____)

(__ × _____) + (__ × _____)

2. Solve using the partial products method.

   Cayla's school has 258 students. Janet's school has 3 times as many students as Cayla's. How many students are in Janet's school?

3. Model with a tape diagram and solve.

   4 times as much as 467

Solve using the standard algorithm, the area model, the distributive property, or the partial products method.

4. $5{,}131 \times 7$

5. 3 times as many as 2,805

6. A restaurant sells 1,725 pounds of spaghetti and 925 pounds of linguini every month. After 9 months, how many pounds of pasta does the restaurant sell?

Name _______________ Date _______________

1. Solve the following expressions using the standard algorithm, the partial products method, and the area model.

a. 302 × 8

8 (300 + 2)

(8 × _____) + (8 × _____)

b. 216 × 5

5 (_____ + _____ + _____)

(__ × _____) + (__ × _____) + (__ × _____)

c. 593 × 9

__ (_____ + _____ + _____)

(__ × _____) + (__ × _____) + (__ × _____)

2. Solve using the partial products method.

   On Monday, 475 people visited the museum. On Saturday, there were 4 times as many visitors as there were on Monday. How many people visited the museum on Saturday?

3. Model with a tape diagram and solve.

   6 times as much as 384

Solve using the standard algorithm, the area model, the distributive property, or the partial products method.

4. 6,253 × 3

5. 7 times as many as 3,073

6. A cafeteria makes 2,516 pounds of white rice and 608 pounds of brown rice every month. After 6 months, how many pounds of rice does the cafeteria make?

4.

a. Write an equation that would allow someone to find the value of R.

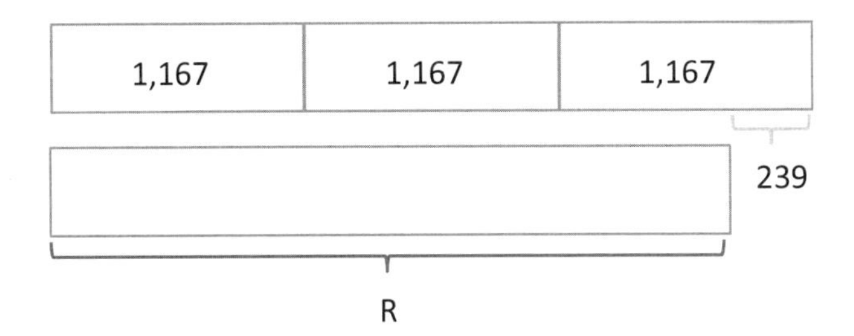

b. Write your own word problem to correspond to the tape diagram, and then solve.

Name ______________________________ Date ______________

Use the RDW process to solve the following problems.

1. The table shows the number of stickers of various types in Chrissy's new sticker book. Chrissy's six friends each own the same sticker book. How many stickers do Chrissy and her six friends have altogether?

Type of Sticker	Number of Stickers
flowers	32
smiley faces	21
hearts	39

2. The small copier makes 437 copies each day. The large copier makes 4 times as many copies each day. How many copies does the large copier make each week?

3. Jared sold 194 Boy Scout chocolate bars. Matthew sold three times as many as Jared. Gary sold 297 fewer than Matthew. How many bars did Gary sell?

4.

a. Write an equation that would allow someone to find the value of M.

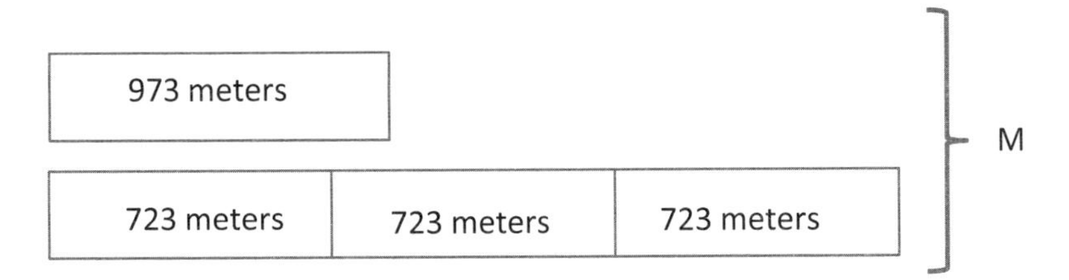

b. Write your own word problem to correspond to the tape diagram, and then solve.

Name ______________________________ Date ________________

Solve using the RDW process.

1. Over the summer, Kate earned $180 each week for 7 weeks. Of that money, she spent $375 on a new computer and $137 on new clothes. How much money did she have left?

2. Sylvia weighed 8 pounds when she was born. By her first birthday, her weight had tripled. By her second birthday, she had gained 12 more pounds. At that time, Sylvia's father weighed 5 times as much as she did. What was Sylvia and her dad's combined weight?

3. Three boxes weighing 128 pounds each and one box weighing 254 pounds were loaded onto the back of an empty truck. A crate of apples was then loaded onto the same truck. If the total weight loaded onto the truck was 2,000 pounds, how much did the crate of apples weigh?

4. In one month, Charlie read 814 pages. In the same month, his mom read 4 times as many pages as Charlie, and that was 143 pages more than Charlie's dad read. What was the total number of pages read by Charlie and his parents?

Name ______________________________ Date ________________

Solve using the RDW process.

1. A pair of jeans costs $89. A jean jacket costs twice as much. What is the total cost of a jean jacket and 4 pairs of jeans?

2. Sarah bought a shirt on sale for $35. The original price of the shirt was 3 times that amount. Sarah also bought a pair of shoes on sale for $28. The original price of the shoes was 5 times that amount. Together, how much money did the shirt and shoes cost before they went on sale?

3. All 3,000 seats in a theater are being replaced. So far, 5 sections of 136 seats and a sixth section containing 348 seats have been replaced. How many more seats do they still need to replace?

4. Computer Depot sold 762 reams of paper. Paper Palace sold 3 times as much paper as Computer Depot and 143 reams more than Office Supply Central. How many reams of paper were sold by all three stores combined?

Name ______________________________ Date ________________

Use the RDW process to solve the following problems.

1. There are 19 identical socks. How many pairs of socks are there? Will there be any socks without a match? If so, how many?

2. If it takes 8 inches of ribbon to make a bow, how many bows can be made from 3 feet of ribbon (1 foot = 12 inches)? Will any ribbon be left over? If so, how much?

3. The library has 27 chairs and 5 tables. If the same number of chairs is placed at each table, how many chairs can be placed at each table? Will there be any extra chairs? If so, how many?

4. The baker has 42 kilograms of flour. She uses 8 kilograms each day. After how many days will she need to buy more flour?

5. Caleb has 76 apples. He wants to bake as many pies as he can. If it takes 8 apples to make each pie, how many apples will he use? How many apples will not be used?

6. Forty-five people are going to the beach. Seven people can ride in each van. How many vans will be required to get everyone to the beach?

Name ______________________________ Date ______________

Use the RDW process to solve the following problems.

1. Linda makes booklets using 2 sheets of paper. She has 17 sheets of paper. How many of these booklets can she make? Will she have any extra paper? How many sheets?

2. Linda uses thread to sew the booklets together. She cuts 6 inches of thread for each booklet. How many booklets can she stitch with 50 inches of thread? Will she have any unused thread after stitching up the booklets? If so, how much?

3. Ms. Rochelle wants to put her 29 students into groups of 6. How many groups of 6 can she make? If she puts any remaining students in a smaller group, how many students will be in that group?

4. A trainer gives his horse, Caballo, 7 gallons of water every day from a 57-gallon container. How many days will Caballo receive his full portion of water from the container? On which number day will the trainer need to refill the container of water?

5. Meliza has 43 toy soldiers. She lines them up in rows of 5 to fight imaginary zombies. How many of these rows can she make? After making as many rows of 5 as she can, she puts the remaining soldiers in the last row. How many soldiers are in that row?

6. Seventy-eight students are separated into groups of 8 for a field trip. How many groups are there? The remaining students form a smaller group of how many students?

Name ______________________________ Date ________________

Show division using an array.	Show division using an area model.
1. 18 ÷ 6 Quotient = ________ Remainder = _______	Can you show 18 ÷ 6 with one rectangle? _____
2. 19 ÷ 6 Quotient = ________ Remainder = _______	Can you show 19 ÷ 6 with one rectangle? _____ Explain how you showed the remainder:

Solve using an array and an area model. The first one is done for you.

Example: 25 ÷ 2

a. 

b. 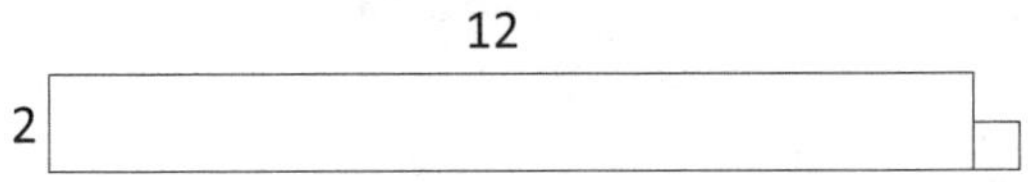

Quotient = 12 Remainder = 1

3. 29 ÷ 3

a.

b.

4. 22 ÷ 5

a.

b.

5. 43 ÷ 4

a.

b.

6. 59 ÷ 7

a.

b.

Name ______________________________ Date ______________

Show division using an array.	Show division using an area model.
1. $24 \div 4$ Quotient = ________ Remainder = _______	Can you show $24 \div 4$ with one rectangle? ______
2. $25 \div 4$ Quotient = ________ Remainder = _______	Can you show $25 \div 4$ with one rectangle? ______ Explain how you showed the remainder:

Solve using an array and area model. The first one is done for you.

Example: 25 ÷ 3

a.

b.

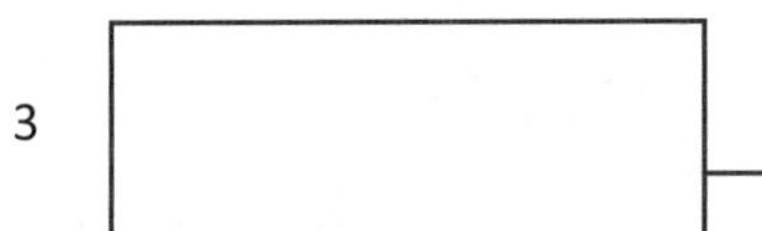

Quotient = 8 Remainder = 1

3. 44 ÷ 7

a.

b.

4. 34 ÷ 6

a.

b.

5. 37 ÷ 6

a.

b.

6. 46 ÷ 8

a.

b.

Name ______________________________ Date ______________

Show the division using disks. Relate your work on the place value chart to long division. Check your quotient and remainder by using multiplication and addition.

1. $7 \div 2$

Ones

$2\overline{)7}$

quotient = ________

remainder = ________

Check Your Work

$$\begin{array}{r} 3 \\ \times\ 2 \\ \hline \end{array}$$

2. $27 \div 2$

Tens	Ones

$2\overline{)27}$

quotient = ________

remainder = ________

Check Your Work

3. $8 \div 3$

Ones

$3\overline{)8}$

quotient = ________

remainder = ________

Check Your Work

4. $38 \div 3$

Tens	Ones

$3\overline{)38}$

quotient = ________

remainder = ________

Check Your Work

5. $6 \div 4$

Ones

$4\overline{)6}$

quotient = ________

remainder = ________

Check Your Work

6. $86 \div 4$

Tens	Ones

$4\overline{)86}$

quotient = ________

remainder = ________

Check Your Work

Name ______________________________ Date ______________

Show the division using disks. Relate your work on the place value chart to long division. Check your quotient and remainder by using multiplication and addition.

1. $7 \div 3$

Ones

$3 \overline{)\,7}$

quotient = ________

remainder = ________

Check Your Work

$$\begin{array}{r} 2 \\ \times\ 3 \\ \hline \end{array}$$

2. $67 \div 3$

Tens	Ones

$3 \overline{)\,67}$

quotient = ________

remainder = ________

Check Your Work

3. $5 \div 2$

Ones

$2 \overline{)\,5}$

quotient = ________

remainder = ________

Check Your Work

4. 85 ÷ 2

Tens	Ones

2 | 85

quotient = ________

remainder = ________

Check Your Work

5. 5 ÷ 4

Ones

4 | 5

quotient = ________

remainder = ________

Check Your Work

6. 85 ÷ 4

Tens	Ones

4 | 85

quotient = ________

remainder = ________

Check Your Work

ones	
tens	

tens place value chart

Name ______________________________ Date ________________

Show the division using disks. Relate your model to long division. Check your quotient and remainder by using multiplication and addition.

1. 5 ÷ 2

Ones

$2\overline{)5}$

quotient = ________

remainder = ________

Check Your Work

$$\begin{array}{r} 2 \\ \times\ 2 \\ \hline \end{array}$$

2. 50 ÷ 2

Tens	Ones

$2\overline{)50}$

quotient = ______

remainder = ______

Check Your Work

3. 7 ÷ 3

Ones

$3\overline{)7}$

quotient = ________

remainder = ________

Check Your Work

4. 75 ÷ 3

Tens	Ones

3 ⟌ 7 5

quotient = ________

remainder = ________

Check Your Work

5. 9 ÷ 4

Ones

4 ⟌ 9

quotient = ________

remainder = ________

Check Your Work

6. 92 ÷ 4

Tens	Ones

4 ⟌ 9 2

quotient = ______

remainder = ______

Check Your Work

Name ______________________________ Date ______________

Show the division using disks. Relate your model to long division. Check your quotient and remainder by using multiplication and addition.

1. $7 \div 2$

Ones

$2\overline{)7}$

quotient = ________

remainder = ________

Check Your Work

2. $73 \div 2$

Tens	Ones

$2\overline{)73}$

quotient = ________

remainder = ________

Check Your Work

3. $6 \div 4$

Ones

$4\overline{)6}$

quotient = ________

remainder = ________

Check Your Work

4. 62 ÷ 4

Tens	Ones

4 ⟌ 6 2

quotient = ______

remainder = ______

Check Your Work

5. 8 ÷ 3

Ones

3 ⟌ 8

quotient = ______

remainder = ______

Check Your Work

6. 84 ÷ 3

Tens	Ones

3 ⟌ 8 4

quotient = ______

remainder = ______

Check Your Work

Name ______________________________ Date ____________________

Solve using the standard algorithm. Check your quotient and remainder by using multiplication and addition.

1. 46 ÷ 2	2. 96 ÷ 3
3. 85 ÷ 5	4. 52 ÷ 4
5. 53 ÷ 3	6. 95 ÷ 4

7. $89 \div 6$	8. $96 \div 6$
9. $60 \div 3$	10. $60 \div 4$
11. $95 \div 8$	12. $95 \div 7$

Name ______________________________ Date ________________

Solve using the standard algorithm. Check your quotient and remainder by using multiplication and addition.

1. $84 \div 2$	2. $84 \div 4$
3. $48 \div 3$	4. $80 \div 5$
5. $79 \div 5$	6. $91 \div 4$

7. $91 \div 6$	8. $91 \div 7$
9. $87 \div 3$	10. $87 \div 6$
11. $94 \div 8$	12. $94 \div 6$

Name ______________________________ Date ______________

1. When you divide 94 by 3, there is a remainder of 1. Model this problem with place value disks. In the place value disk model, how did you show the remainder?

2. Cayman says that 94 ÷ 3 is 30 with a remainder of 4. He reasons this is correct because (3 × 30) + 4 = 94. What mistake has Cayman made? Explain how he can correct his work.

3. The place value disk model is showing 72 ÷ 3. Complete the model. Explain what happens to the 1 ten that is remaining in the tens column.

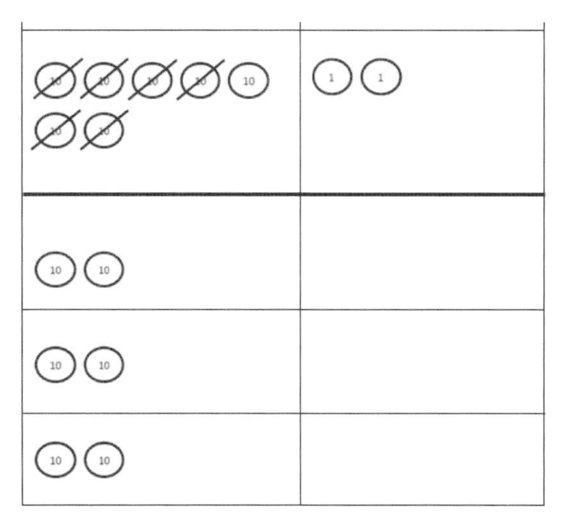

4. Two friends evenly share 56 dollars.
   a. They have 5 ten dollar bills and 6 one dollar bills. Draw a picture to show how the bills will be shared. Will they have to make change at any stage?

   b. Explain how they share the money evenly.

5. Imagine you are filming a video explaining the problem $45 \div 3$ to new fourth graders. Create a script to explain how you can keep dividing after getting a remainder of 1 ten in the first step.

Name ______________________________ Date ________________

1. When you divide 86 by 4, there is a remainder of 2. Model this problem with place value disks. In the place value disk model, how can you see that there is a remainder?

2. Francine says that 86 ÷ 4 is 20 with a remainder of 6. She reasons this is correct because (4 × 20) + 6 = 86. What mistake has Francine made? Explain how she can correct her work.

3. The place value disk model is showing 67 ÷ 4. Complete the model. Explain what happens to the 2 tens that are remaining in the tens column.

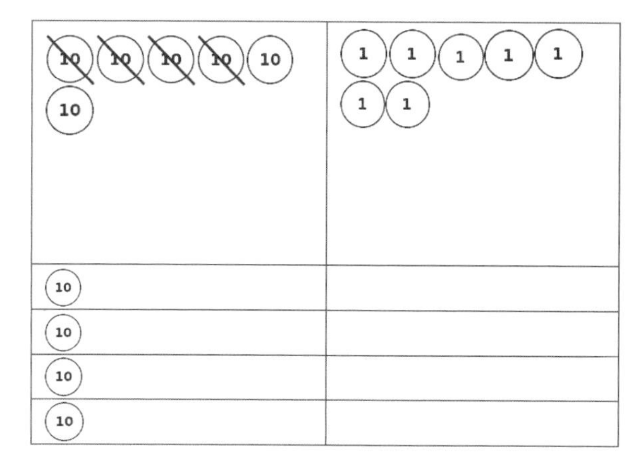

4. Two friends share 76 blueberries.
   a. To count the blueberries, they put them into small bowls of 10 blueberries. Draw a picture to show how the blueberries can be shared equally. Will they have to split apart any of the bowls of 10 blueberries when they share them?

   b. Explain how the friends can share the blueberries fairly.

5. Imagine you are drawing a comic strip showing how to solve the problem $72 \div 4$ to new fourth graders. Create a script to explain how you can keep dividing after getting a remainder of 3 tens in the first step.

Name ______________________________ Date ________________

1. Alfonso solved a division problem by drawing an area model.
   a. Look at the area model. What division problem did Alfonso solve?

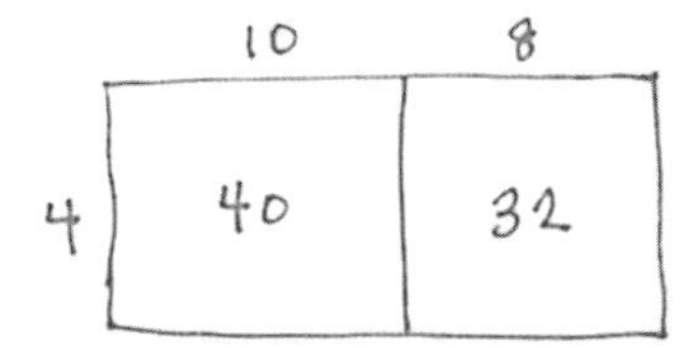

   b. Show a number bond to represent Alfonso's area model. Start with the total and then show how the total is split into two parts. Below the two parts, represent the total length using the distributive property and then solve.

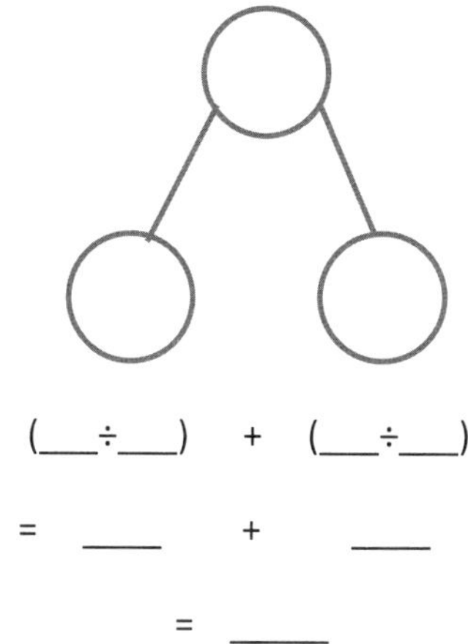

(___÷___) + (___÷___)

= ____ + ____

= _____

2. Solve 45 ÷ 3 using an area model. Draw a number bond and use the distributive property to solve for the unknown length.

3. Solve $64 \div 4$ using an area model. Draw a number bond to show how you partitioned the area, and represent the division with a written method.

4. Solve $92 \div 4$ using an area model. Explain, using words, pictures, or numbers, the connection of the distributive property to the area model.

5. Solve $72 \div 6$ using an area model and the standard algorithm.

Name ______________________________ Date ______________

1. Maria solved the following division problem by drawing an area model.

   a. Look at the area model. What division problem did Maria solve?

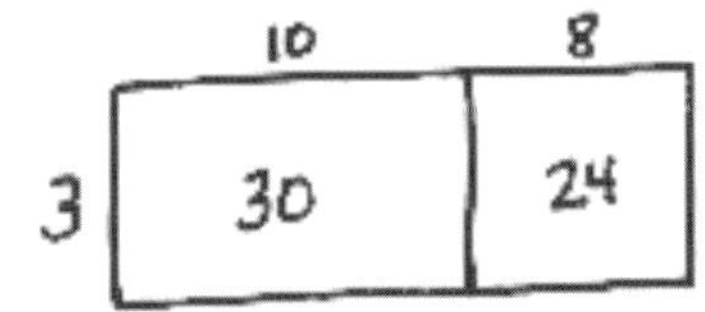

   b. Show a number bond to represent Maria's area model. Start with the total and then show how the total is split into two parts. Below the two parts, represent the total length using the distributive property and then solve.

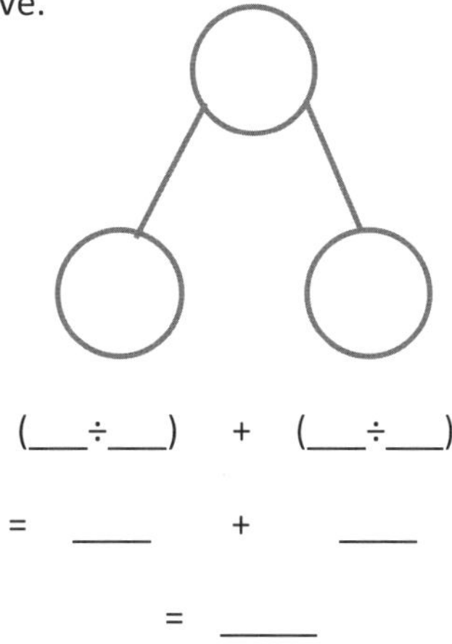

(___÷___) + (___÷___)

= _____ + _____

= _____

2. Solve $42 \div 3$ using an area model. Draw a number bond and use the distributive property to solve for the unknown length.

3. Solve $60 \div 4$ using an area model. Draw a number bond to show how you partitioned the area, and represent the division with a written method.

4. Solve $72 \div 4$ using an area model. Explain, using words, pictures, or numbers, the connection of the distributive property to the area model.

5. Solve $96 \div 6$ using an area model and the standard algorithm.

Name ________________________________ Date ________________

1. Solve 37 ÷ 2 using an area model. Use long division and the distributive property to record your work.

2. Solve 76 ÷ 3 using an area model. Use long division and the distributive property to record your work.

3. Carolina solved the following division problem by drawing an area model.

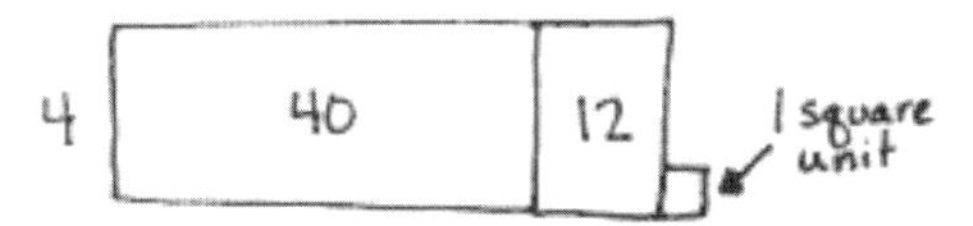

a. What division problem did she solve?

b. Show how Carolina's model can be represented using the distributive property.

Solve the following problems using the area model. Support the area model with long division or the distributive property.

4. $48 \div 3$	5. $49 \div 3$
6. $56 \div 4$	7. $58 \div 4$
8. $66 \div 5$	9. $79 \div 3$

10. Seventy-three students are divided into groups of 6 students each. How many groups of 6 students are there? How many students will not be in a group of 6?

Name ___________________________ Date _______________

1. Solve 35 ÷ 2 using an area model. Use long division and the distributive property to record your work.

2. Solve 79 ÷ 3 using an area model. Use long division and the distributive property to record your work.

3. Paulina solved the following division problem by drawing an area model.

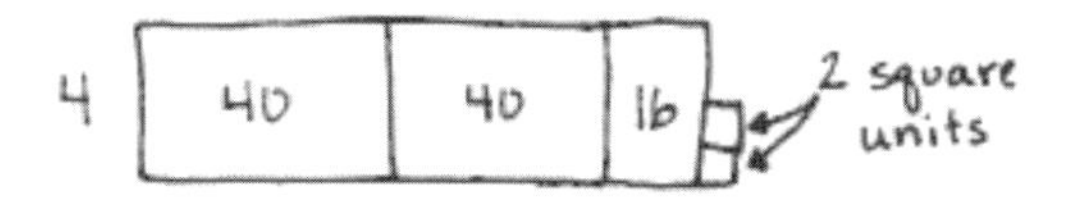

a. What division problem did she solve?
b. Show how Paulina's model can be represented using the distributive property.

Solve the following problems using the area model. Support the area model with long division or the distributive property.

4. $42 \div 3$	5. $43 \div 3$
6. $52 \div 4$	7. $54 \div 4$
8. $61 \div 5$	9. $73 \div 3$

10. Ninety-seven lunch trays were placed equally in 4 stacks. How many lunch trays were in each stack? How many lunch trays will be left over?

Name ______________________________ Date ________________

1. Record the factors of the given numbers as multiplication sentences and as a list in order from least to greatest. Classify each as prime (P) or composite (C). The first problem is done for you.

	Multiplication Sentences	Factors	P or C
a.	4 $1 \times 4 = 4$ $2 \times 2 = 4$	The factors of 4 are: 1, 2, 4	C
b.	6	The factors of 6 are:	
c.	7	The factors of 7 are:	
d.	9	The factors of 9 are:	
e.	12	The factors of 12 are:	
f.	13	The factors of 13 are:	
g.	15	The factors of 15 are:	
h.	16	The factors of 16 are:	
i.	18	The factors of 18 are:	
j.	19	The factors of 19 are:	
k.	21	The factors of 21 are:	
l.	24	The factors of 24 are:	

2. Find all factors for the following numbers, and classify each number as prime or composite. Explain your classification of each as prime or composite.

Factor Pairs for 25	

Factor Pairs for 28	

Factor Pairs for 29	

3. Bryan says all prime numbers are odd numbers.
   a. List all of the prime numbers less than 20 in numerical order.

   b. Use your list to show that Bryan's claim is false.

4. Sheila has 28 stickers to divide evenly among 3 friends. She thinks there will be no leftovers. Use what you know about factor pairs to explain if Sheila is correct.

Name ______________________________ Date ______________

1. Record the factors of the given numbers as multiplication sentences and as a list in order from least to greatest. Classify each as prime (P) or composite (C). The first problem is done for you.

	Multiplication Sentences	Factors	P or C
a.	8 $1 \times 4 = 8$ $2 \times 4 = 8$	The factors of 8 are: 1, 2, 4, 8	C
b.	10	The factors of 10 are:	
c.	11	The factors of 11 are:	
d.	14	The factors of 14 are:	
e.	17	The factors of 17 are:	
f.	20	The factors of 20 are:	
g.	22	The factors of 22 are:	
h.	23	The factors of 23 are:	
i.	25	The factors of 25 are:	
j.	26	The factors of 26 are:	
k.	27	The factors of 27 are:	
l.	28	The factors of 28 are:	

2. Find all factors for the following numbers, and classify each number as prime or composite. Explain your classification of each as prime or composite.

Factor Pairs for 19	

Factor Pairs for 21	

Factor Pairs for 24	

3. Bryan says that only even numbers are composite.
   a. List all of the odd numbers less than 20 in numerical order.

   b. Use your list to show that Bryan's claim is false.

4. Julie has 27 grapes to divide evenly among 3 friends. She thinks there will be no leftovers. Use what you know about factor pairs to explain whether or not Julie is correct.

Name ______________________________ Date ______________

1. Explain your thinking or use division to answer the following.

a. Is 2 a factor of 84?	b. Is 2 a factor of 83?
c. Is 3 a factor of 84?	d. Is 2 a factor of 92?
e. Is 6 a factor of 84?	f. Is 4 a factor of 92?
g. Is 5 a factor of 84?	h. Is 8 a factor of 92?

2. Use the associative property to find more factors of 24 and 36.

a. $24 = 12 \times 2$

$= (\_\_\_ \times 3) \times 2$

$= \_\_\_ \times (3 \times 2)$

$= \_\_\_ \times 6$

$= \_\_\_$

b. $36 = \_\_\_\_ \times 4$

$= (\_\_\_\_ \times 3) \times 4$

$= \_\_\_\_ \times (3 \times 4)$

$= \_\_\_\_ \times 12$

$= \_\_\_\_$

3. In class, we used the associative property to show that when 6 is a factor, then 2 and 3 are factors, because $6 = 2 \times 3$. Use the fact that $8 = 4 \times 2$ to show that 2 and 4 are factors of 56, 72, and 80.

$56 = 8 \times 7$ $72 = 8 \times 9$ $80 = 8 \times 10$

4. The first statement is false. The second statement is true. Explain why, using words, pictures, or numbers.

If a number has 2 and 4 as factors, then it has 8 as a factor.

If a number has 8 as a factor, then both 2 and 4 are factors.

Name ______________________________ Date ______________

1. Explain your thinking or use division to answer the following.

a. Is 2 a factor of 72?	b. Is 2 a factor of 73?
c. Is 3 a factor of 72?	d. Is 2 a factor of 60?
e. Is 6 a factor of 72?	f. Is 4 a factor of 60?
g. Is 5 a factor of 72?	h. Is 8 a factor of 60?

2. Use the associative property to find more factors of 12 and 30.

a. $12 = 6 \times 2$

$= (\_\_\_ \times 2) \times 2$

$= \_\_\_ \times (2 \times 2)$

$= \_\_\_ \times \_\_\_$

$= \_\_\_$

b. $30 = \_\_\_\_ \times 5$

$= (\_\_\_\_ \times 3) \times 5$

$= \_\_\_\_ \times (3 \times 5)$

$= \_\_\_\_ \times 15$

$= \_\_\_\_$

3. In class, we used the associative property to show that when 6 is a factor, then 2 and 3 are factors, because $6 = 2 \times 3$. Use the fact that $10 = 5 \times 2$ to show that 2 and 5 are factors of 70, 80, and 90.

$70 = 10 \times 7$ $\quad$ $80 = 10 \times 8$ $\quad$ $90 = 10 \times 9$

4. The first statement is false. The second statement is true. Explain why, using words, pictures, or numbers.

If a number has 2 and 6 as factors, then it has 12 as a factor.

If a number has 12 as a factor, then both 2 and 6 are factors.

Name ______________________________ Date ________________

1. For each of the following, time yourself for 1 minute. See how many multiples you can write.

   a. Write the multiples of 5 starting from 100.

   b. Write the multiples of 4 starting from 20.

   c. Write the multiples of 6 starting from 36.

2. List the numbers that have 24 as a multiple.

3. Use mental math, division, or the associative property to solve. (Use scratch paper if you like.)

   a. Is 12 a multiple of 4? ______ Is 4 a factor of 12? _______

   b. Is 42 a multiple of 8? ______ Is 8 a factor of 42? _______

   c. Is 84 a multiple of 6? ______ Is 6 a factor of 84? _______

4. Can a prime number be a multiple of any other number except itself? Explain why or why not.

5. Follow the directions below.

1	2	3	4	5	6	7	8	9	10
11	12	13	14	15	16	17	18	19	20
21	22	23	24	25	26	27	28	29	30
31	32	33	34	35	36	37	38	39	40
41	42	43	44	45	46	47	48	49	50
51	52	53	54	55	56	57	58	59	60
61	62	63	64	65	66	67	68	69	70
71	72	73	74	75	76	77	78	79	80
81	82	83	84	85	86	87	88	89	90
91	92	93	94	95	96	97	98	99	100

a. Circle in red the multiples of 2. When a number is a multiple of 2, what are the possible values for the ones digit?

b. Shade in green the multiples of 3. Choose one. What do you notice about the sum of the digits? Choose another. What do you notice about the sum of the digits?

c. Circle in blue the multiples of 5. When a number is a multiple of 5, what are the possible values for the ones digit?

d. Draw an X over the multiples of 10. What digit do all multiples of 10 have in common?

Name ______________________________ Date ________________

1. For each of the following, time yourself for 1 minute. See how many multiples you can write.

   a. Write the multiples of 5 starting from 75.

   b. Write the multiples of 4 starting from 40.

   c. Write the multiples of 6 starting from 24.

2. List the numbers that have 30 as a multiple.

3. Use mental math, division, or the associative property to solve. (Use scratch paper if you like.)

   a. Is 12 a multiple of 3? ______ Is 3 a factor of 12? _______

   b. Is 48 a multiple of 8? ______ Is 48 a factor of 8? _______

   c. Is 56 a multiple of 6? ______ Is 6 a factor of 56? _______

4. Can a prime number be a multiple of any other number except itself? Explain why or why not.

5. Follow the directions below.

1	2	3	4	5	6	7	8	9	10
11	12	13	14	15	16	17	18	19	20
21	22	23	24	25	26	27	28	29	30
31	32	33	34	35	36	37	38	39	40
41	42	43	44	45	46	47	48	49	50
51	52	53	54	55	56	57	58	59	60
61	62	63	64	65	66	67	68	69	70
71	72	73	74	75	76	77	78	79	80
81	82	83	84	85	86	87	88	89	90
91	92	93	94	95	96	97	98	99	100

a. Underline the multiples of 6. When a number is a multiple of 6, what are the possible values for the ones digit?

b. Draw a square around the multiples of 4. Look at the multiples of 4 that have an odd number in the tens place. What values do they have in the ones place?

c. Look at the multiples of 4 that have an even number in the tens place. What values do they have in the ones place? Do you think this pattern would continue with multiples of 4 that are larger than 100?

d. Circle the multiples of 9. Choose one. What do you notice about the sum of the digits? Choose another one. What do you notice about the sum of the digits?

Name ______________________________ Date ______________

1. Follow the directions.

   Shade the number 1 red.

   a. Circle the first unmarked number.
   b. Cross off every multiple of that number except the one you circled. If it's already crossed off, skip it.
   c. Repeat Steps (a) and (b) until every number is either circled or crossed off.
   d. Shade every crossed out number in orange.

1	2	3	4	5	6	7	8	9	10
11	12	13	14	15	16	17	18	19	20
21	22	23	24	25	26	27	28	29	30
31	32	33	34	35	36	37	38	39	40
41	42	43	44	45	46	47	48	49	50
51	52	53	54	55	56	57	58	59	60
61	62	63	64	65	66	67	68	69	70
71	72	73	74	75	76	77	78	79	80
81	82	83	84	85	86	87	88	89	90
91	92	93	94	95	96	97	98	99	100

2.

a. List the circled numbers.

b. Why were the circled numbers not crossed off along the way?

c. Except for the number 1, what is similar about all of the numbers that were crossed off?

d. What is similar about all of the numbers that were circled?

Name ______________________________ Date ________________

1. A student used the sieve of Eratosthenes to find all prime numbers less than 100. Create a step-by-step set of directions to show how it was completed. Use the word bank to help guide your thinking as you write the directions. Some words may be used just once, more than once, or not at all.

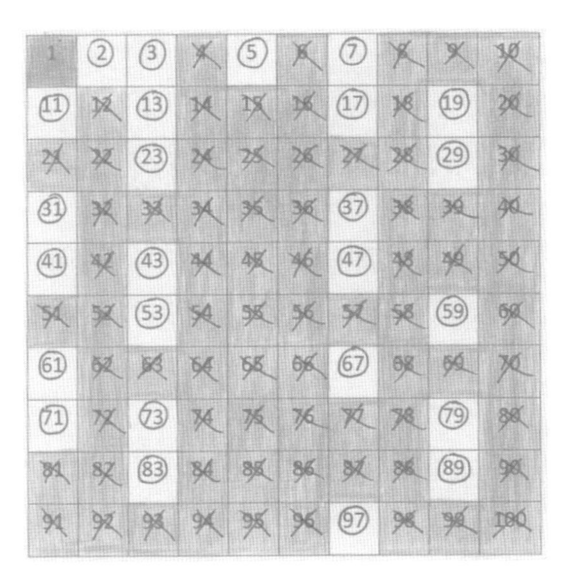

Word Bank

composite	cross out
number	shade
circle	X
multiple	prime

Directions for completing the sieve of Eratosthenes activity:

2. What do all of the numbers that are crossed out have in common?

3. What do all of the circled numbers have in common?

4. There is one number that is neither crossed out nor circled. Why is it treated differently?

Name _______________________________ Date _______________

1. Draw place value disks to represent the following problems. Rewrite each in unit form and solve.

   a. 6 ÷ 2 = ________

      6 ones ÷ 2 = ________ ones

      (1) (1) (1)   (1) (1) (1)

   b. 60 ÷ 2 = ________

      6 tens ÷ 2 = ____________

   c. 600 ÷ 2 = ________

      ______________ ÷ 2 = ______________

   d. 6,000 ÷ 2 = ________

      ______________ ÷ 2 = ______________

2. Draw place value disks to represent each problem. Rewrite each in unit form and solve.

   a. 12 ÷ 3 = ________

      12 ones ÷ 3 = ________ ones

   b. 120 ÷ 3 = ________

      ______________________ ÷ 3 = ______________________

   c. 1,200 ÷ 3 = ________

      ______________________ ÷ 3 = ______________________

3. Solve for the quotient. Rewrite each in unit form.

a. 800 ÷ 2 = 400 8 hundreds ÷ 2 = 4 hundreds	b. 600 ÷ 2 = ________	c. 800 ÷ 4 = ________	d. 900 ÷ 3 = ________
e. 300 ÷ 6 = ________ 30 tens ÷ 6 = ____ tens	f. 240 ÷ 4 = ________	g. 450 ÷ 5 = ________	h. 200 ÷ 5 = ________
i. 3,600 ÷ 4 = ________ 36 hundreds ÷ 4 = ____ hundreds	j. 2,400 ÷ 4 = ________	k. 2,400 ÷ 3 = ________	l. 4,000 ÷ 5 = ______

4. Some sand weighs 2,800 kilograms. It is divided equally among 4 trucks. How many kilograms of sand are in each truck?

5. Ivy has 5 times as many stickers as Adrian has. Ivy has 350 stickers. How many stickers does Adrian have?

6. An ice cream stand sold $1,600 worth of ice cream on Saturday, which was 4 times the amount sold on Friday. How much money did the ice cream stand collect on Friday?

Name ______________________________ Date ______________

1. Draw place value disks to represent the following problems. Rewrite each in unit form and solve.

   a. 6 ÷ 3 = ________

      6 ones ÷ 3 = ________ ones

      (1)(1) (1)(1) (1)(1)

   b. 60 ÷ 3 = ________

      6 tens ÷ 3 = ____________

   c. 600 ÷ 3 = ________

      ______________ ÷ 3 = ______________

   d. 6,000 ÷ 3 = ________

      ______________ ÷ 3 = ______________

2. Draw place value disks to represent each problem. Rewrite each in unit form and solve.

   a. 12 ÷ 4 = ________

      12 ones ÷ 4 = ________ ones

   b. 120 ÷ 4 = ________

      ______________________ ÷ 4 = ______________________

   c. 1,200 ÷ 4 = ________

      ______________________ ÷ 4 = ______________________

3. Solve for the quotient. Rewrite each in unit form.

a. 800 ÷ 4 = 200 8 hundreds ÷ 4 = 2 hundreds	b. 900 ÷ 3 = ________	c. 400 ÷ 2 = ________	d. 300 ÷ 3 = ________
e. 200 ÷ 4 = ________ 20 tens ÷ 4 = ____ tens	f. 160 ÷ 2 = ________	g. 400 ÷ 5 = ________	h. 300 ÷ 5 = ________
i. 1,200 ÷ 3 = ________ 12 hundreds ÷ 3 = ____ hundreds	j. 1,600 ÷ 4 = ________	k. 2,400 ÷ 4 = ________	l. 3,000 ÷ 5 = ________

4. A fleet of five fire engines carries a total of 20,000 liters of water. If each truck holds the same amount of water, how many liters of water does each truck carry?

5. Jamie drank 4 times as much juice as Brodie. Jamie drank 280 milliliters of juice. How much juice did Brodie drink?

6. A diner sold $2,400 worth of French fries in June, which was 4 times as much as was sold in May. How many dollars' worth of French fries were sold at the diner in May?

thousands	hundreds	tens	ones

thousands place value chart for dividing

Name ______________________________ Date ______________

1. Divide. Use place value disks to model each problem.

a. 324 ÷ 2	b. 344 ÷ 2
c. 483 ÷ 3	d. 549 ÷ 3

2. Model using place value disks and record using the algorithm.

a. 655 ÷ 5	
Disks	Algorithm
b. 726÷ 3	
Disks	Algorithm
c. 688 ÷ 4	
Disks	Algorithm

Name ______________________________ Date ______________

1. Divide. Use place value disks to model each problem.

a. 346 ÷ 2	b. 528 ÷ 2
c. 516 ÷ 3	d. 729 ÷ 3

2. Model using place value disks, and record using the algorithm.

a. $648 \div 4$ Disks	Algorithm
b. $755 \div 5$ Disks	Algorithm
c. $964 \div 4$ Disks	Algorithm

Name ___________________________________ Date ____________________

1. Divide. Check your work by multiplying. Draw disks on a place value chart as needed.

a. 574 ÷ 2	b. 861 ÷ 3
c. 354 ÷ 2	d. 354 ÷ 3
e. 873 ÷ 4	f. 591 ÷ 5

g. 275 ÷ 3	h. 459 ÷ 5
i. 678 ÷ 4	j. 955 ÷ 4

2. Zach filled 581 one-liter bottles with apple cider. He distributed the bottles to 4 stores. Each store received the same number of bottles. How many liter bottles did each of the stores receive? Were there any bottles left over? If so, how many?

Name ______________________ Date ______________

1. Divide. Check your work by multiplying. Draw disks on a place value chart as needed.

a. 378 ÷ 2	b. 795 ÷ 3
c. 512 ÷ 4	d. 492 ÷ 4
e. 539 ÷ 3	f. 862 ÷ 5

g. 498 ÷ 3	h. 783 ÷ 5
i. 621 ÷ 4	j. 531 ÷ 4

2. Selena's dog completed an obstacle course that was 932 meters long. There were 4 parts to the course, all equal in length. How long was 1 part of the course?

Name ______________________________ Date ________________

1. Divide, and then check using multiplication.

a. 1,672 ÷ 4	b. 1,578 ÷ 4
c. 6,948 ÷ 2	d. 8,949 ÷ 4
 e. 7,569 ÷ 2	 f. 7,569 ÷ 3

g. 7,955 ÷ 5	h. 7,574 ÷ 5
i. 7,469 ÷ 3	j. 9,956 ÷ 4

2. There are twice as many cows as goats on a farm. All the cows and goats have a total of 1,116 legs. How many goats are there?

Name ______________________________ Date ______________

1. Divide, and then check using multiplication.

a. 2,464 ÷ 4	b. 1,848 ÷ 3
c. 9,426 ÷ 3	d. 6,587 ÷ 2
e. 5,445 ÷ 3	f. 5,425 ÷ 2

g. 8,467 ÷ 3	h. 8,456 ÷ 3
i. 4,937 ÷ 4	j. 6,173 ÷ 5

2. A truck has 4 crates of apples. Each crate has an equal number of apples. Altogether, the truck is carrying 1,728 apples. How many apples are in 3 crates?

Name ______________________________ Date ________________

Divide. Check your solutions by multiplying.

1. $204 \div 4$

2. $704 \div 3$

3. $627 \div 3$

4. $407 \div 2$

5. $760 \div 4$

6. $5{,}120 \div 4$

7. 3,070 ÷ 5

8. 6,706 ÷ 5

9. 8,313 ÷ 4

10. 9,008 ÷ 3

11.

a. Find the quotient and remainder for 3,131 ÷ 3.

b. How could you change the digit in the ones place of the whole so that there would be no remainder? Explain how you determined your answer.

Name ____ Date ______________________

Divide. Check your solutions by multiplying.

1. $409 \div 5$

2. $503 \div 2$

3. $831 \div 4$

4. $602 \div 3$

5. $720 \div 3$

6. $6{,}250 \div 5$

7. $2,060 \div 5$

8. $9,031 \div 2$

9. $6,218 \div 4$

10. $8,000 \div 4$

Name _______________________________ Date _______________

Draw a tape diagram and solve. The first two tape diagrams have been drawn for you. Identify if the group size or the number of groups is unknown.

1. Monique needs exactly 4 plates on each table for the banquet. If she has 312 plates, how many tables is she able to prepare?

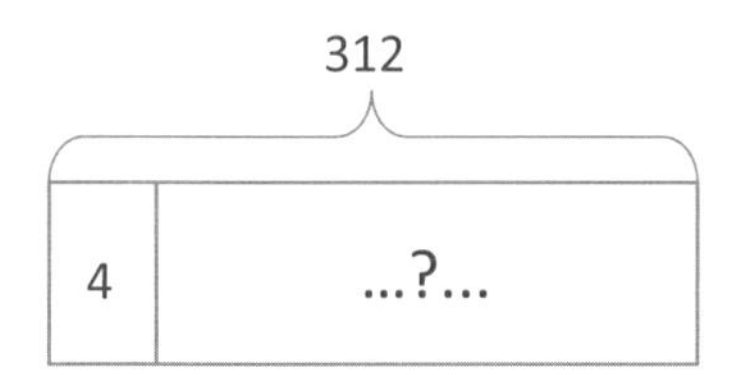

2. 2,365 books were donated to an elementary school. If 5 classrooms shared the books equally, how many books did each class receive?

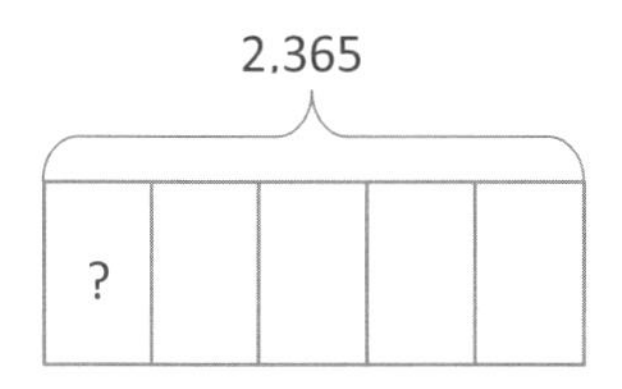

3. If 1,503 kilograms of rice was packed in sacks weighing 3 kilograms each, how many sacks were packed?

4. Rita made 5 batches of cookies. There was a total of 2,400 cookies. If each batch contained the same number of cookies, how many cookies were in 4 batches?

5. Every day, Sarah drives the same distance to work and back home. If Sarah drove 1,005 miles in 5 days, how far did Sarah drive in 3 days?

Name ______________________________ Date ________________

Solve the following problems. Draw tape diagrams to help you solve. Identify if the group size or the number of groups is unknown.

1. 500 milliliters of juice was shared equally by 4 children. How many milliliters of juice did each child get?

2. Kelly separated 618 cookies into baggies. Each baggie contained 3 cookies. How many baggies of cookies did Kelly make?

3. Jeff biked the same distance each day for 5 days. If he traveled 350 miles altogether, how many miles did he travel each day?

4. A piece of ribbon 876 inches long was cut by a machine into 4-inch long strips to be made into bows. How many strips were cut?

5. Five Martians equally share 1,940 Groblarx fruits. How many Groblarx fruits will 3 of the Martians receive?

Name ______________________________ Date ______________

Solve the following problems. Draw tape diagrams to help you solve. If there is a remainder, shade in a small portion of the tape diagram to represent that portion of the whole.

1. A concert hall contains 8 sections of seats with the same number of seats in each section. If there are 248 seats, how many seats are in each section?

2. In one day, the bakery made 719 bagels. The bagels were divided into 9 equal shipments. A few bagels were left over and given to the baker. How many bagels did the baker get?

3. The sweet shop has 614 pieces of candy. They packed the candy into bags with 7 pieces in each bag. How many bags of candy did they fill? How many pieces of candy were left?

4. There were 904 children signed up for the relay race. If there were 6 children on each team, how many teams were made? The remaining children served as referees. How many children served as referees?

5. 1,188 kilograms of rice are divided into 7 sacks. How many kilograms of rice are in 6 sacks of rice? How many kilograms of rice remain?

Name ______________________________ Date ________________

Solve the following problems. Draw tape diagrams to help you solve. If there is a remainder, shade in a small portion of the tape diagram to represent that portion of the whole.

1. Meneca bought a package of 435 party favors to give to the guests at her birthday party. She calculated that she could give 9 party favors to each guest. How many guests is she expecting?

2. 4,000 pencils were donated to an elementary school. If 8 classrooms shared the pencils equally, how many pencils did each class receive?

3. 2,008 kilograms of potatoes were packed into sacks weighing 8 kilograms each. How many sacks were packed?

4. A baker made 7 batches of muffins. There was a total of 252 muffins. If there was the same number of muffins in each batch, how many muffins were in a batch?

5. Samantha ran 3,003 meters in 7 days. If she ran the same distance each day, how far did Samantha run in 3 days?

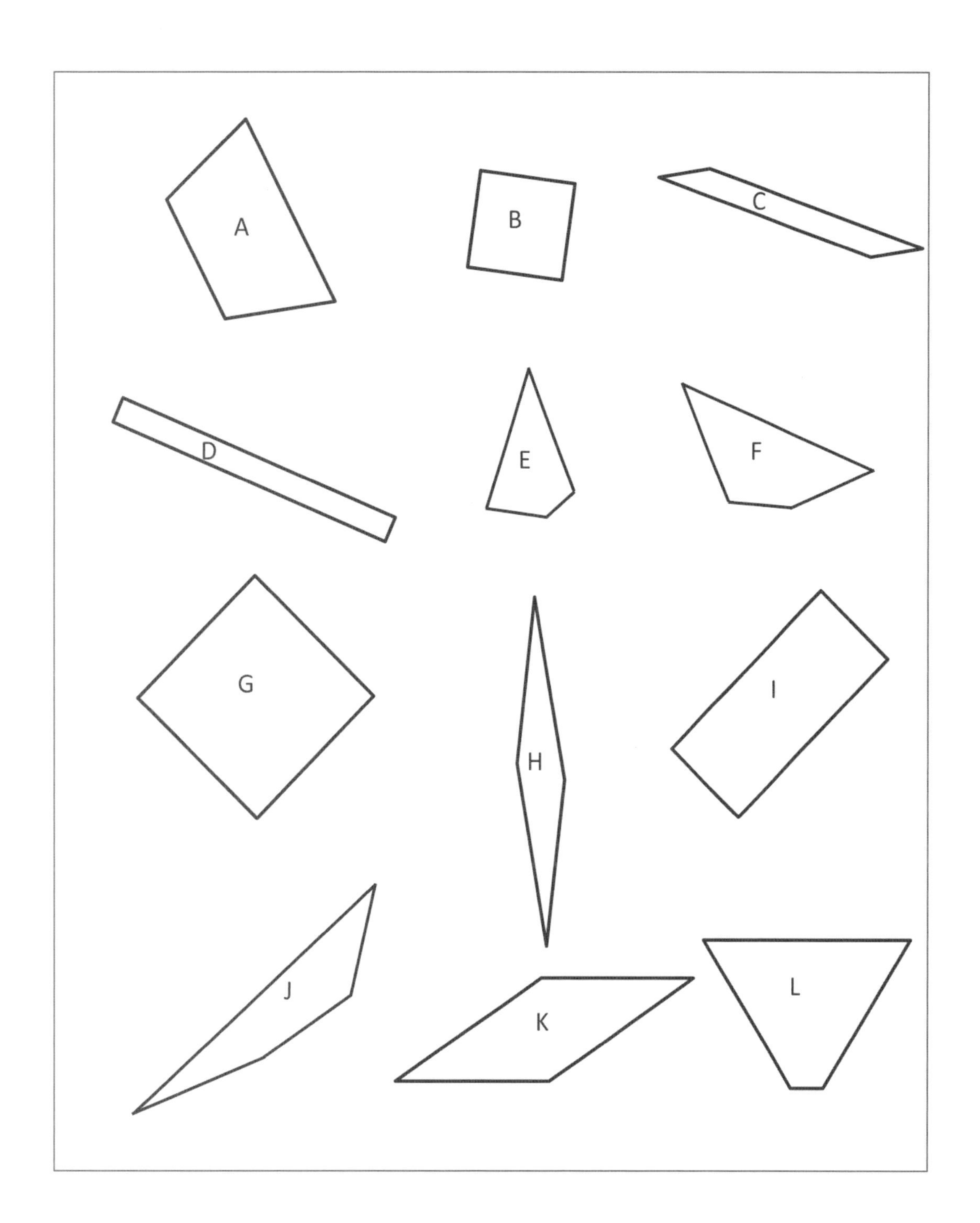

shapes

3. a. Draw an area model to solve 774 ÷ 3.

   b. Draw a number bond to represent this problem.

   c. Record your work using the long division algorithm.

4. a. Draw an area model to solve 1,584 ÷ 2.

   b. Draw a number bond to represent this problem.

   c. Record your work using the long division algorithm.

Name ______________________________ Date ______________

1. Arabelle solved the following division problem by drawing an area model.

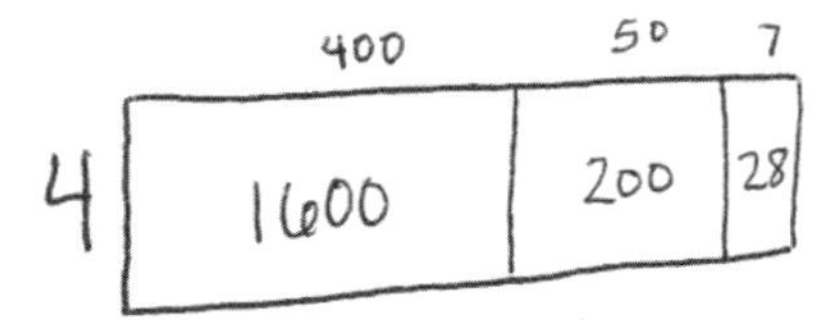

   a. What division problem did she solve?

   b. Show a number bond to represent Arabelle's area model, and represent the total length using the distributive property.

2. a. Solve $816 \div 4$ using the area model. There is no remainder in this problem.

   b. Draw a number bond and use a written method to record your work from (a).

3. a. Draw an area model to solve 549 ÷ 3.

b. Draw a number bond to represent this problem.

c. Record your work using the long division algorithm.

4. a. Draw an area model to solve 2,762 ÷ 2.

b. Draw a number bond to represent this problem.

c. Record your work using the long division algorithm.

Name ______________________________ Date ________________

1. Use the associative property to rewrite each expression. Solve using disks, and then complete the number sentences.

a. 30 × 24

= (____ × 10) × 24

= ____ × (10 × 24)

= ______

hundreds	tens	ones

b. 40 × 43

= (4 × 10) × _____

= 4 × (10 × ____)

= ______

thousands	hundreds	tens	ones

c. 30 × 37

= (3 × ____) × _____

= 3 × (10 × _____)

= ______

thousands	hundreds	tens	ones

2. Use the associative property and place value disks to solve.

   a. 20 × 27

   b. 40 × 31

3. Use the associative property without place value disks to solve.

   a. 40 × 34

   b. 50 × 43

4. Use the distributive property to solve the following problems. Distribute the second factor.

   a. 40 × 34

   b. 60 × 25

Name ______________________________ Date ________________

1. Use the associative property to rewrite each expression. Solve using disks, and then complete the number sentences.

a. 20 × 34

= (____ × 10) × 34

= ____ × (10 × 34)

= _______

hundreds	tens	ones

b. 30 × 34

= (3 × 10) × _____

= 3 × (10 × ___)

= _______

thousands	hundreds	tens	ones

c. 30 × 42

= (3 × ____) × _____

= 3 × (10 × _____)

= _______

thousands	hundreds	tens	ones

2. Use the associative property and place value disks to solve.

   a. 20×16

   b. 40×32

3. Use the associative property without place value disks to solve.

   a. 30×21

   b. 60×42

4. Use the distributive property to solve the following. Distribute the second factor.

   a. 40×43

   b. 70×23

Name ______________________ Date ____________

Use an area model to represent the following expressions. Then, record the partial products and solve.

1. 20×22

$$\begin{array}{r} 22 \\ \times\ 20 \\ \hline \\ +\quad \\ \hline \end{array}$$

2. 50×41

$$\begin{array}{r} 41 \\ \times\ 50 \\ \hline \\ +\quad \\ \hline \end{array}$$

3. 60×73

$$\begin{array}{r} 73 \\ \times\ 60 \\ \hline \\ +\quad \\ \hline \end{array}$$

Draw an area model to represent the following expressions. Then, record the partial products vertically and solve.

4. 80×32

5. 70×54

Visualize the area model and solve the following expressions numerically.

6. 30×68

7. 60×34

8. 40×55

9. 80×55

Name ______________________________ Date ______________

Use an area model to represent the following expressions. Then, record the partial products and solve.

1. 30 × 17

$$\begin{array}{r} 17 \\ \times\ 30 \\ \hline \\ +\quad \\ \hline \end{array}$$

2. 40 × 58

$$\begin{array}{r} 58 \\ \times\ 40 \\ \hline \\ +\quad \\ \hline \end{array}$$

3. 50 × 38

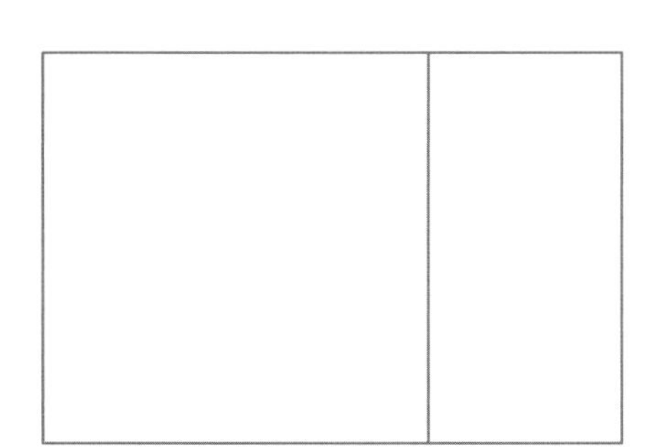

$$\begin{array}{r} 38 \\ \times\ 50 \\ \hline \\ +\quad \\ \hline \end{array}$$

Draw an area model to represent the following expressions. Then, record the partial products vertically and solve.

4. 60×19

5. 20×44

Visualize the area model and solve the following expressions numerically.

6. 20×88

7. 30×88

8. 70×47

9. 80×65

Name ______________________________ Date ______________

1.

a. In each of the two models pictured below, write the expressions that determine the area of each of the four smaller rectangles.

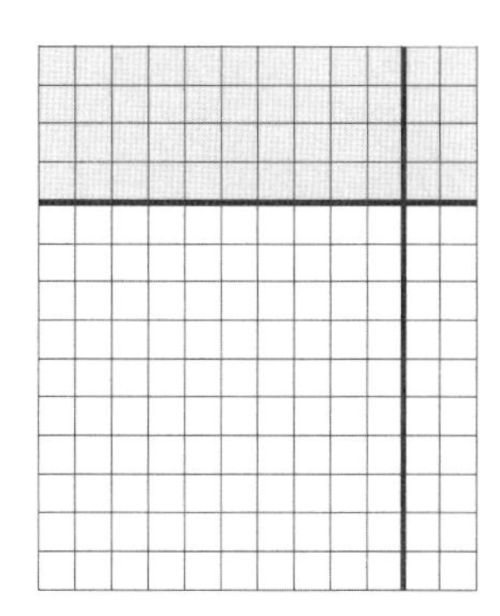

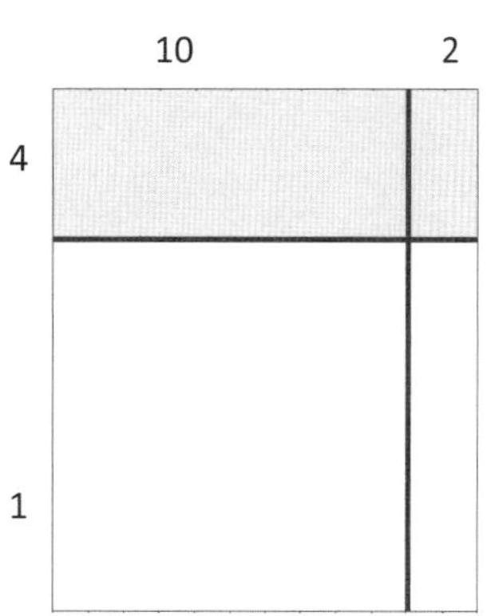

b. Using the distributive property, rewrite the area of the large rectangle as the sum of the areas of the four smaller rectangles. Express first in number form and then read in unit form.

$14 \times 12 =$ (4 × _____) + (4 × _____) + (10 × _____) + (10 × _____)

2. Use an area model to represent the following expression. Record the partial products and solve.

a. 14 × 22

$$\begin{array}{r} 22 \\ \times\ 14 \\ \hline \_\_\_ \\ \_\_\_ \\ \_\_\_ \\ +\ \_\_\_ \\ \hline \end{array}$$

Draw an area model to represent the following expressions. Record the partial products vertically and solve.

3. 25×32

4. 35×42

Visualize the area model and solve the following numerically using four partial products. (You may sketch an area model if it helps.)

5. 42×11

6. 46×11

Name ______________________________ Date ______________

1.

a. In each of the two models pictured below, write the expressions that determine the area of each of the four smaller rectangles.

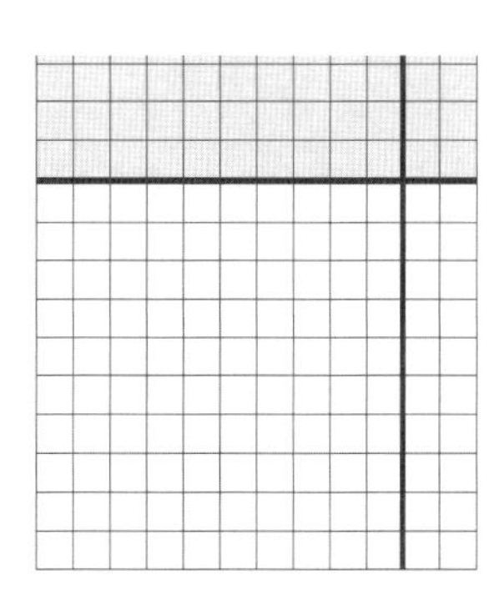

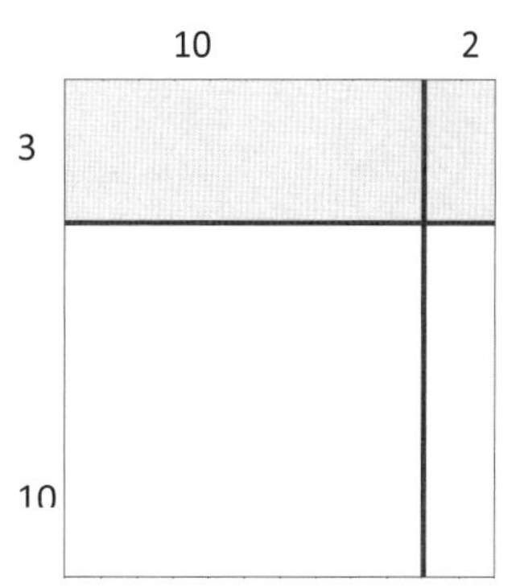

b. Using the distributive property, rewrite the area of the large rectangle as the sum of the areas of the four smaller rectangles. Express first in number form and then read in unit form.

13 × 12 = (3 × _____) + (3 × _____) + (10 × _____) + (10 × _____)

Use an area model to represent the following expression. Record the partial products and solve.

2. 17 × 34

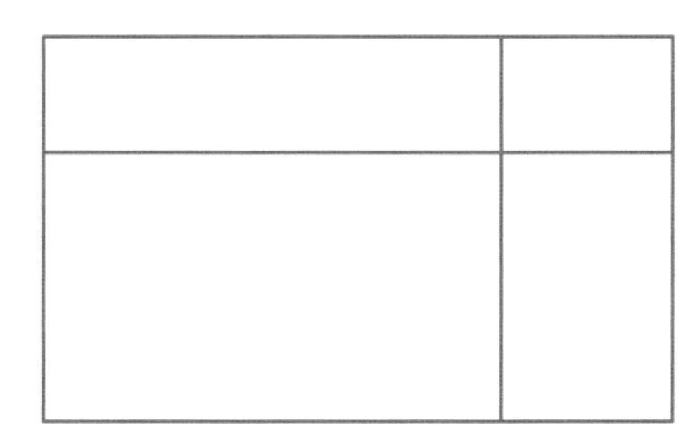

$$\begin{array}{r} 3\ 4 \\ \times\ 1\ 7 \\ \hline \_\_\_ \\ \_\_\_ \\ +\ \_\_\_ \\ \hline \end{array}$$

Draw an area model to represent the following expressions. Record the partial products vertically and solve.

3. 45×18

4. 45×19

Visualize the area model and solve the following numerically using four partial products. (You may sketch an area model if it helps.)

5. 12×47

6. 23×93

7. 23×11

8. 23×22

Name ______________________________ Date ______________

1. Solve 14 × 12 using 4 partial products and 2 partial products. Remember to think in terms of units as you solve. Write an expression to find the area of each smaller rectangle in the area model.

10 2

4

10

$$\begin{array}{r} 1\ 2 \\ \times\ 1\ 4 \\ \hline \end{array}$$

_______ *4 ones × 2 ones*

_______ *4 ones × 1 ten*

_______ *1 ten × 2 ones*

_______ *1 ten × 1 ten*

12

4

10

$$\begin{array}{r} 1\ 2 \\ \times\ 1\ 4 \\ \hline \end{array}$$

_______ *4 ones × 12 ones*

_______ *1 ten × 12 ones*

2. Solve 32 × 43 using 4 partial products and 2 partial products. Match each partial product to its area on the models. Remember to think in terms of units as you solve.

40 3

2

30

$$\begin{array}{r} 4\ 3 \\ \times\ 3\ 2 \\ \hline \end{array}$$

_______ *2 ones × 3 ones*

_______ *2 ones × 4 tens*

_______ *3 tens × 3 ones*

_______ *3 tens × 4 tens*

43

2

30

$$\begin{array}{r} 4\ 3 \\ \times\ 3\ 2 \\ \hline \end{array}$$

_______ *2 ones × 43 ones*

_______ *3 tens × 43 ones*

3. Solve 57 × 15 using 2 partial products. Match each partial product to its rectangle on the area model.

4. Solve the following using 2 partial products. Visualize the area model to help you.

a.
```
   2 5
 × 4 6
 -----
        ___ × ___
 -----
        ___ × ___
 -----
```

b.
```
   1 8
 × 6 2
 -----
        ___ × ___
 -----
        ___ × ___
 -----
```

c.
```
   3 9
 × 4 6
 -----
```

d.
```
   7 8
 × 2 3
 -----
```

Name ___________________________ Date ______________

1. Solve 26 × 34 using 4 partial products and 2 partial products. Remember to think in terms of units as you solve. Write an expression to find the area of each smaller rectangle in the area model.

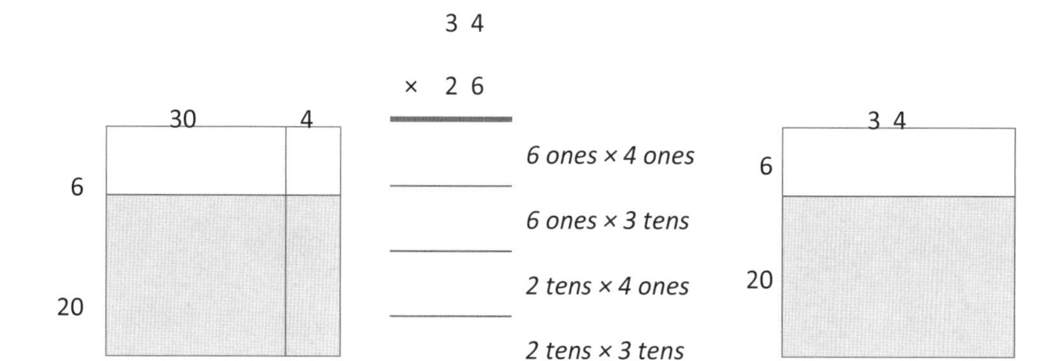

2. Solve using 4 partial products and 2 partial products. Remember to think in terms of units as you solve. Write an expression to find the area of each smaller rectangle in the area model.

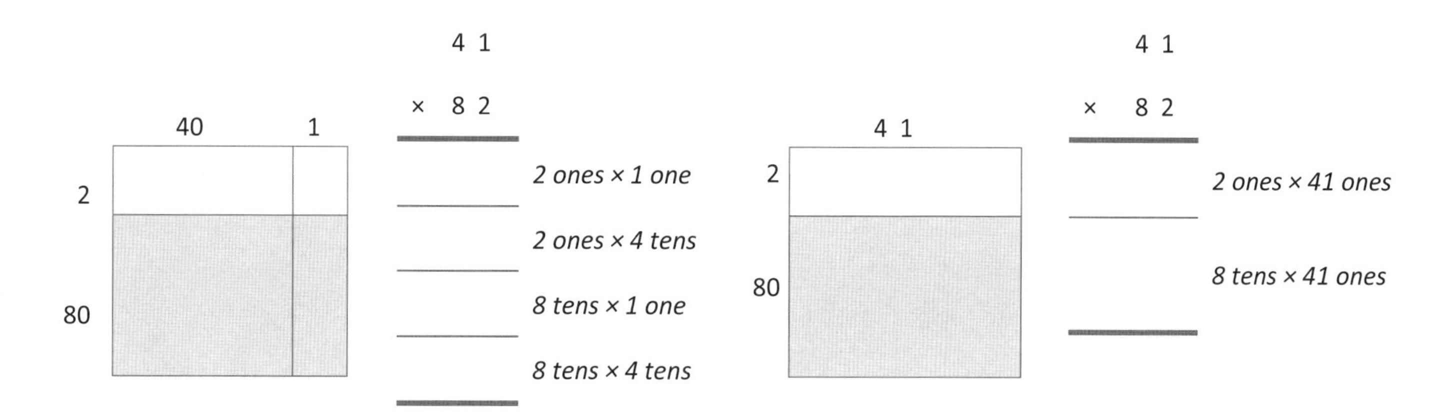

3. Solve 52 × 26 using 2 partial products and an area model. Match each partial product to its area on the model.

4. Solve the following using 2 partial products. Visualize the area model to help you.

a.
```
    6 8
  × 2 3
  -----
          ___ × ___
  -----
          ___ × ___
  -----
```

b.
```
    4 9
  × 3 3
  -----
          ___ × ___
  -----
          ___ × ___
  -----
```

c.
```
    1 6
  × 2 5
  -----
```

d.
```
    5 4
  × 7 1
  -----
```

Name ______________________________ Date ______________

1. Express 23 × 54 as two partial products using the distributive property. Solve.

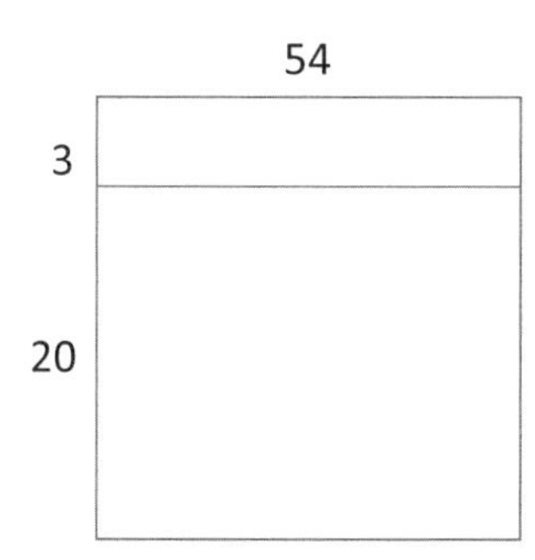

23 × 54 = (___ fifty-fours) + (___ fifty-fours)

```
   5 4
 × 2 3
 -----
 _____   3 × _____
 _____  20 × _____
```

2. Express 46 × 54 as two partial products using the distributive property. Solve.

54
6
40

46 × 54 = (___ fifty-fours) + (___ fifty-fours)

```
   5 4
 × 4 6
 -----
 _____   _____ × _____
 _____   _____ × _____
```

3. Express 55 × 47 as two partial products using the distributive property. Solve.

55 × 47 = (____ × _____) + (____ × _____)

```
   4 7
 × 5 5
 -----
 _____   _____ × _____
 _____   _____ × _____
```

4. Solve the following using 2 partial products.

```
      5 8
×     4 5
---------
            ____ × ____
---------
            ____ × ____
---------
```

5. Solve using the multiplication algorithm.

```
      8 2
×     5 5
---------
            ____ × ____
---------
            ____ × ____
---------
```

6. 53 × 63

7. 84 × 73

Name ______________________________ Date ______________

1. Express 26 × 43 as two partial products using the distributive property. Solve.

43

6

20

26 × 43 = (_____ forty-threes) + (_____ forty-threes)

```
    4 3
×   2 6
-------
_______   6 × _____
_______   20 × _____
```

2. Express 47 × 63 as two partial products using the distributive property. Solve.

47 × 63 = (_____ sixty-threes) + (_____ sixty-threes)

```
    6 3
×   4 7
-------
_______   _____ × _____
_______   _____ × _____
```

3. Express 54 × 67 as two partial products using the distributive property. Solve.

54 × 67 = (___ × ____) + (___ × ____)

```
    6 7
×   5 4
-------
_______   _____ × _____
_______   _____ × _____
```

4. Solve the following using two partial products.

$$\begin{array}{r} 5\ 2 \\ \times\ \ 3\ 4 \\ \hline \end{array}$$

____ × ____

____ × ____

5. Solve using the multiplication algorithm.

$$\begin{array}{r} 8\ 6 \\ \times\ \ 5\ 6 \\ \hline \end{array}$$

____ × ____

____ × ____

6. 54 × 52

7. 44 × 76

8. 63 × 63

9. 68 × 79

Notes

Notes